An Intuitive Primer on Effective
Functional Genomics Study Design

Yoav Gilad

I

Published by Yoav Gilad
Chicago, IL

First Edition: January 2025

ISBN: 979-8-218-58595-2
Library of Congress Control Number: 2025927727

For permissions requests, please contact: *yoav.gilad@gmail.com*

Thank you to my colleagues who read earlier versions of this primer and provided valuable comments and suggestions. I am also grateful to the students I have taught over the years, who helped me refine my own understanding of study design. And to my kids, I apologize. I know you were hoping that when I finally wrote a book, it would be colorful and filled with more pictures.

Reader, I hope you find this book useful. If you come across any typos or errors, have suggestions for improvements, additional topics to include, or simply want to share your thoughts about the book, feel free to reach out to me at yoav.gilad@gmail.com

To Doron, my mentor

Table of Contents

A personal note

In 2006, I developed and started teaching a course on functional genomics. This was a relatively new field at the time. My course focused on introducing students to different genomic technologies (back in 2006, mainly different types of microarrays), different protocols for data collection, the principles of effective study design, and a little bit about data structures and the framework for multivariate statistical analysis.

I continued to teach this course until 2023, and it evolved significantly over the years. In 2008, I added a section on the then-newer ("next-generation") sequencing technologies, and gradually, the course shifted to focus more on sequencing data, eventually phasing out mention of microarrays altogether. The course was further updated to include cutting-edge functional genomics protocols and data types, relevant newer experimental models like stem cells, related techniques like genome editing, and most recently, single-cell and spatial transcriptomic data.

What I realized, is that throughout all these changes to the course, the unit on effective study design has remained nearly unchanged. In my course, we discuss foundational aspects of study design in a dedicated unit that is being taught for about three to four hours. We continuously mention and further discuss aspects of study design throughout the course, and these have evolved as the specific topics we discuss have changed over the years; still, the unit explicitly dedicated to the discussion of the foundational principles of effective functional genomics study design has remained mostly the same.

Our understanding of study designs has continued to evolve since 2006, of course. New genomic technologies, new data structures, and new and more effective analysis approaches have all contributed and necessitated further advancement in study designs. Yet, the basic foundational principles of effective study design have not changed, and the unit dedicated to study design in my course has provided many of the students with their first introduction to these enduring principles.

The impetus for developing the course, and ultimately for writing this primer, stemmed from the recognition that data collection, and therefore study design, must come before data analysis. While this statement is trivial, it is also – I believe – the basis for a behavior observed among many scientists and students, namely, that data collection often begins without a deep understanding of the issues relevant to data analysis or study design. This happens because many scientists prioritize data collection (it comes first!), and often turn to learning about data analysis only when it becomes immediately relevant, usually after the data have been gathered. Consequently, formal education on the intricacies of study design is also often acquired later. This common conduct can lead to critical errors in the data collection pipeline that cannot be corrected.

Similarly, when researchers in empirical fields aim to collaborate with computational or statistical experts, they often approach these experts only after completing their experimental work. Unfortunately, errors made in the study design by empirical labs, before they engaged with their computational/statistician collaborators, are often irreparable.

The foundational understanding of study design provided in the course and, hopefully, in this primer, is intended to bridge this gap by offering an accessible resource that requires no prior knowledge of math or statistics. The goal is to encourage early, thoughtful planning and to foster an appreciation for the key principles that support effective functional genomics experimentation. My aim was to create a text that feels approachable and can be read at any time. I wanted to avoid creating a resource that readers might feel the need to set aside, thinking it requires too much time, focus, or additional background to fully engage with. I also wanted to avoid writing a book that readers feel they must postpone engaging with until they "have more time" - a moment that often comes only after their data collection is already complete.

Because this primer is designed to be approachable and focused, it does not aim to provide a comprehensive review of all topics related to study design. What we are going to discuss here are principles and examples that provide basic intuition for the properties that are important to consider in order to achieve effective functional genomics study design. We do not use much math or statistics, and even discussions of variance components are kept to a qualitative – almost intuitive – level. My approach is very much influenced by the book *'Calculus for Cats'* by Kenn Amdahl and Jim Loats. That book transformed my own intuitive understanding of calculus when I first read it. My goal when I developed the genomics course was to try and imitate the approach (though, regrettably without cats) and explain the intuition behind the principles of study design with as few equations as possible. To do this, I tried to use precise language in the course and avoid the pitfalls of general descriptions.

While students did not always think that my course was overall outstanding (I have detailed feedback notes…), I can say that they always appreciated the unit on study design. I finally decided to try and put it in writing, and I hope that this primer will be found useful.

A final note on the way I wrote this primer: In my course, I have always enjoyed exploring answers to different questions by encouraging students to propose and refine their own solutions through discussion. This has consistently been one of my favorite aspects of teaching, as it sparks lively and thought-provoking conversations. For many students, these discussions provide their first opportunity to deeply consider the specifics and implications of nuanced decisions in study design.

To bring some of that experience into this primer, I have included moments where you are encouraged to pause and reflect on a question before reading further. I believe that engaging with these questions on your own, even if your answer is not entirely correct, helps prepare your mind to understand the concepts more deeply. By thinking through the problem first, you will already be considering the key ideas, making it easier to connect the information that follows and complete the picture.

Why do we need a book specifically focused on designing functional genomics studies?

Most study designs in biology aim to detect **specific** *deviations from expectations, targeting particular outcomes that align closely with the hypothesis being tested. By contrast, in functional genomics, the study design must accommodate the search for* **any** *deviation from expectations, as the goal is often to uncover unexpected patterns across a vast, high-dimensional dataset.*

Study design is a fundamental aspect of any scientific experiment. Regardless of the field, a well-designed study ensures that conclusions are based on valid data and that confounding factors do not obscure the true relationships being investigated. However, in functional genomics, the nature of the data, and the questions being asked, create unique challenges, amplifying the importance of addressing confounders and biases with great care.

To understand this, it helps to contrast functional genomics with more traditional types of experiments. In classical biology, hypotheses are often specific and tied directly to a measurable outcome. For example, a researcher might hypothesize that "providing more water to a plant will result in increased fruit size." This hypothesis is quite specific, and it leads naturally to a straightforward experimental design: set up a control group and an experimental group (where additional water is provided to the latter), measure fruit size, and ensure that other factors such as soil quality or sunlight are consistent. The measured outcome is clearly defined, and confounders can often be directly controlled.

Functional genomics, by contrast, operates on an entirely different scale. Genomic studies often address broad, unspecific questions, such as "will this treatment cause changes in gene expression?" Such an aim does not specify which genes or pathways will be affected or how. Instead of focusing on a single outcome, the goal is to uncover patterns across tens of thousands of measurements, such as identifying a set (dozens, hundreds, thousands?) of differentially expressed genes between conditions.

This shift from measuring specific outcomes to the exploration of broad patterns introduces a central challenge that lies at the heart of study design in functional genomics. On the one hand, the richness of the data allows researchers to uncover system-wide patterns and gain insights that would be impossible to achieve in traditional studies. On the other hand, this same richness makes functional genomics studies particularly vulnerable to confounders and biases, which can profoundly influence the results while − counterintuitively - often remain difficult to detect.

In functional genomics, the way we interpret results is fundamentally different from more traditional experiments. Instead of focusing on a single outcome, like whether watering a plant makes its fruit grow larger, we analyze tens of thousands of outcomes simultaneously. In gene regulatory studies, for example, each gene follows its own pattern of expression, and most of the time, these patterns fall within what we expect, similar to how most runners finish within a predictable range of times in a race.

What makes a result of a genomic study interesting is when a gene shows an unexpected pattern, behaving differently from what we'd predict based on the overall data. Any gene that stands out can become the focus of our attention, as its behavior might reveal something meaningful about the biological processes at work. This flexibility to find "interesting" results across thousands of genes is one of the strengths of functional genomics but also one of its vulnerabilities.

Small errors or biases in the experiment, confounders, can subtly influence the data, causing some genes to stand out for reasons unrelated to the conditions, treatments, or perturbations that are the focus of the study. In hypothesis-driven studies, confounders can also affect results, but their impact is often easier to identify because there are only one or a few outcomes to analyze. In genomics, however, the high dimensionality of the data and the tendency to consider "any unusual pattern" worthy of further study, make the effects of confounders both significant and elusive. A cryptic confounder in functional genomics studies might alter thousands of measurements in subtle ways, resulting in distorted patterns that appear statistically robust but are biologically quite meaningless.

Among the many excellent texts on study design, *Experimental Design for the Life Sciences* by Graeme Ruxton and Nick Colegrave stands out to me as a clear and comprehensive resource. However, while it offers invaluable guidance on the principles of experimental design in biology, its broader focus includes only limited attention to the specific challenges of functional genomics, such as managing batch effects, collecting and interpreting high-dimensional data, and addressing the subtle but profound influence of confounders.

This primer aims to fill that gap by addressing the specific needs of functional genomics. It offers practical guidance for designing studies that are robust, reproducible, and capable of generating reliable insights. The emphasis is on careful planning to tackle the unique challenges of functional genomics: harnessing the richness of high-dimensional data while mitigating the hidden risks of biases and confounders. By emphasizing these foundational approaches, this primer provides a versatile framework to support diverse research efforts in functional genomics.

While the examples in this book primarily focus on gene expression data, the principles described are relevant across the spectrum of functional genomics data types. Challenges related to experimental design, variability, and confounding factors are common to all genomics studies, making the strategies outlined here applicable to other data types such as chromatin accessibility, DNA methylation, or proteomics. By adopting a thoughtful approach to study design, researchers can ensure that their results are both statistically sound and biologically meaningful.

Beyond our scope: Key study design topics not covered

The primer is written in a way that emphasizes the development of intuitive understanding and appreciation for the principles of effective study design, focusing on functional genomics, and staying away from formal mathematical treatments. As a result, several specialized topics that are relevant to study design are **beyond the scope of this text**. While these areas are critical, they require either deeper treatment or a level of formalism that this primer does not aim to provide. Examples of essential topics we do not discuss in this primer include:

- *Standard formulations of classic study designs*. This primer does not include explanations for standard study designs such as one-factorial designs, two-factorial designs, experiments with multiple dependent variables, or distinctions between correlation and manipulation. Additionally, designs like Latin squares, crossover studies, and longitudinal study designs, among others, are not covered. Readers seeking foundational understanding and examples of these topics are encouraged to consult other textbooks or chapters on general study design.
- *Power analysis and sample size estimation*. This primer does not cover the statistical methodologies needed to calculate sample size or perform a power analysis. Readers

are encouraged to consult additional resources for guidance on these critical planning steps (a few recommendations are provided at the end of this primer).

- *Data integration across omics in study design.* The complexities of designing studies that integrate multiple omics datasets (e.g., proteomics, transcriptomics) are not addressed here. Discussing the design of such studies is challenging without exploring the analytical approaches that drive specific design choices.

- *Specific protocols and laboratory workflows.* This primer focuses on general principles of study design rather than the specific details of sample processing and laboratory protocols, which are also critical to consider. For example, while we discuss the study design implications of single-cell experiments, we do not explore workflow-specific details such as cell isolation techniques or handling sparsity.

- *Ethical considerations in study design.* While ethics is a crucial component of study design, this primer does not address topics such as obtaining informed consent, ensuring data privacy, or adhering to regulations governing research involving human or animal subjects. Readers are strongly encouraged to consult resources dedicated to research ethics for guidance on these important aspects.

This primer serves as an entry point into the principles of study design, emphasizing how careful planning directly contributes to a study's success. By maintaining a focused scope and avoiding formal mathematical treatments, it aims to deliver clear and accessible guidance, offering a practical foundation for effective experimental design in functional genomics.

Who is this primer for?

This primer is written with a specific audience in mind: upper-level undergraduate students majoring in biology and first-year graduate students beginning their journey in research. At the same time, it holds relevance for researchers and educators who wish to reinforce their understanding of foundational study design principles in functional genomics and related experimental fields.

For upper-level undergraduates, the primer serves as a bridge between theoretical knowledge and practical application. These students may have completed foundational coursework in molecular biology but are often unfamiliar with the principles of study design that ensure experiments are thoughtfully planned and executed. This text introduces critical concepts, such as defining research goals, identifying confounders, and balancing precision with generalizability, in a manner that is accessible and intuitive.

For first-year graduate students, this primer offers support as they transition from coursework to independent research. The shift to designing their own studies often presents a steep learning curve, and this primer equips them with tools to critically evaluate experimental setups, troubleshoot potential pitfalls, and design experiments with reproducibility and impact in mind. It focuses on essential concepts like replication strategies, managing biological variability, and mitigating confounders, helping students lay a solid foundation for their research projects. It is also a practical resource when preparing pilot studies or navigating collaborative projects where study design decisions influence downstream analyses.

I hope that this primer will also be valuable for educators and researchers seeking to enhance their own teaching or revisit the fundamentals of experimental planning. Educators can use it as a resource in courses or workshops on experimental design, genomics, or related areas. The intuitive approach complements more formal training in statistics or computational biology, offering a way to introduce principles without requiring a heavy mathematical focus. Researchers, particularly those working in empirical fields, may find the primer useful for revisiting foundational concepts that can be overlooked in practice. It highlights the importance of integrating study design considerations early in the research process and provides practical examples to address common study design challenges.

Ultimately, this primer aims to address a critical gap in understanding. In my experience, while empirical researchers often focus on mastering data collection techniques or analytical methods, study design tends to receive less attention. This imbalance can lead to fundamental issues that compromise the validity or interpretability of results, often in ways that cannot be corrected after the fact. By offering language, intuition, and practical examples, I hope that this primer can serve as a starting point for anyone involved in genomic research where study design plays a pivotal role. This primer does not replace formal instruction in statistical or mathematical frameworks but provides an approachable foundation for building confidence in this essential aspect of research.

What is a confounder?

A **confounder** is a variable associated with both the independent variable (the variable being investigated) and the dependent variable (the measured outcome), potentially distorting their observed relationship. This dual association can make it seem like one variable influences the other when the effect may be entirely due to the confounder. For example, in a study of the relationship between genetic variants and gene expression, tissue composition could act as a confounder. If tissue composition is linked to both a genetic variant and the expression of certain genes, failing to account for this association might lead to an incorrect conclusion about the genetic variant's direct role in regulating gene expression.

In genomics studies, confounders frequently arise through batch effects, which occur when systematic differences during sample processing, such as variation in equipment, reagents, or personnel, create patterns in the data. For example, if samples from different tissues are processed in separate laboratories, differences in sample processing can lead to a confounding batch effect.

This primer focuses on strategies to eliminate or minimize confounders through careful and deliberate study design. It is important to acknowledge, however, that not all confounders can be measured or identified. Unmeasured confounders, which often cannot be directly identified or accounted for, are the primary reason functional genomics studies can uncover correlations but cannot definitively establish causation. For instance, an inferred direct link between a genetic variant and gene expression might be influenced by an unmeasured factor, such as DNA methylation. Confounders are an inherent challenge in genomics, but thoughtful study design can help mitigate their impact and enhance reliability, ultimately leading to more robust and interpretable results.

The most important misconception about study design

One of the most pervasive and dangerous misconceptions about study design is the belief that any study can be "saved" through post-hoc methods like data cleaning, advanced statistical analysis, or other manipulations. Many researchers assume that while poor study design may reduce statistical power or increase the difficulty of achieving robust results, it cannot render a study entirely useless. This assumption is not only incorrect, but it also undermines the very foundation of scientific inference.

Poor design can render a study irreparable, incapable of producing valid insights.

Poor design can absolutely invalidate a study, making it impossible to draw meaningful or unambiguous conclusions. When key variables are confounded, biases are introduced, or critical factors are neglected, the data collected may no longer have bearing on the question the study was intended to answer. In such cases, no amount of statistical wizardry or retrospective adjustments can rescue the study from its flawed foundation.

Consider a hypothetical study designed to compare gene expression between tumor and healthy tissue samples. In this study, tumor samples are processed immediately after extraction, while healthy tissue samples are left at room temperature for an extended period before being processed. In this scenario, RNA degradation becomes a confounding variable, as RNA tends to degrade when left at room temperature for prolonged periods. Consequently, RNA from healthy tissues will be systematically more degraded than RNA from tumor samples. This systematic difference in RNA quality means that any observed differences in gene expression between tumors and healthy tissues could reflect true biological variation, differences in RNA quality, or a combination of both.

Because the experimental design failed to account for this confounder, no clear or unambiguous conclusions can be drawn about the biological differences between healthy and tumor samples. Even worse, the confounder is inherent to the data, making it impossible for statistical analysis or data cleaning to disentangle the effects of RNA degradation from those of tissue type (healthy vs. tumor). As a result, the collected data are entirely unusable for addressing the original research question.

This problem is not limited to biological studies. In any field, poor design can introduce systematic biases that fundamentally undermine the validity of a study. It is essential to recognize that poor study design does not simply make results weaker or less precise. It can make them entirely meaningless. A well-designed study ensures that the data collected are appropriate for answering the research question and that potential confounders are avoided, minimized, or explicitly accounted for. Without these safeguards, researchers risk wasting time, resources, and effort on studies that fail to produce reliable insights, or worse, publishing misleading results and conclusions.

The most important takeaway is this: good study design is not optional. It is the foundation on which all subsequent analysis and interpretation rest. A flawed design cannot be fixed after the fact, and no amount of data manipulation can salvage a study that has failed to address the fundamental principles of experimental rigor. To produce reliable, interpretable results, we must prioritize thoughtful, deliberate design from the very beginning.

The first task is to define the goal of the study

Don't approach a functional genomics study with the sole aim of collecting data. Data collection is the journey, not the destination; it is a tool, a means to achieve deeper biological understanding.

Each year, when teaching the Genomics class, I begin the discussion on study design by asking students what should be the first step before considering a functional genomics study design, and I ask for examples. This question is often met with hesitation, leaving the room unusually quiet, even in years when the class is otherwise highly interactive.

The answer to the first part of my question is straightforward: **a functional genomics study should be designed to address a specific goal.** When we discuss effective study designs, the term "effective" refers to the ability to successfully answer questions, test hypotheses, and / or gain meaningful insights. That said, the students' hesitation is not unwarranted. While they immediately appreciate that every study design must address a clear goal, they also intuitively recognize that defining the aims of functional genomics studies can be both challenging and confusing, which makes it difficult for them to provide clear examples.

To explain the students' confusion, it may be helpful to begin with a brief discussion about setting the goals of more traditional studies in biology. The way we articulate scientific hypotheses in casual conversation is rarely precise or detailed enough to fully define the corresponding study design. Nevertheless, in traditional hypothesis-driven biological studies, even a casually described hypothesis often provides enough intuition to convey the essence of the study. For example, stating the hypothesis that "feeding fish a high-protein diet will increase their growth rate" is simple and direct, providing much of the basis for designing a focused experiment. This simple hypothesis **directly connects the intervention (diet) with a measurable outcome (growth rate).**

To design the study and properly test the hypothesis that high-protein diet leads to bigger fish, it is necessary to move beyond a casual statement and provide greater specificity in defining the goals of the study. To do so effectively, we must carefully delineate the scope of the inference we aim to make and identify the range of potential alternative explanations we wish to rule out. The scope determines how broadly the inference can be generalized, and given the scope, we can identify which alternative explanations need to be explicitly addressed and excluded.

When we asked the question regarding the impact of a high-protein diet on fish growth rate, key details were not yet specified, such as the species of fish, life stage, environmental conditions, population density, and dietary protein levels. These specifics are critical because they define the scope of the study. We might, for instance, aim to investigate how increasing dietary protein content from 20 percent to 40 percent affects the growth rate of juvenile zebrafish under controlled tank conditions. This refined focus establishes the scope of the study.

To gain this specific insight, we would need to design a study that effectively rules out alternative explanations for changes in fish growth rate. **Given the specificity of our hypothesis, we can concentrate on a defined set of relevant factors and potential confounders, rather than accounting for every possible variable.** For instance, we might control for water quality, fish density in the tank, feeding schedules, fish age, sex, and genetic variability. *Keep this example of a limited list of potential confounders in mind*; it will take on a different significance when we consider the goals of exploratory functional genomics studies.

Defining the scope of a study requires careful consideration, striking a balance between the specificity needed for clear, actionable conclusions and the generality required for broader insights. Although achieving this balance might seem challenging at first, hypothesis-driven studies typically fall into a few well-established structural categories, such as case-control, time-course, or causal inference studies. These categories provide a framework for defining the goals of the study, its scope, and quite often, a clear range of potential confounders to address. Once the principles of balancing specificity and generality are understood, they can be applied readily across most configurations.

Many of the same principles apply to the design of functional genomics studies, but there is also a profound difference: **functional genomics studies are not often driven by the need to test specific hypotheses, and therefore the list of potential confounders is much harder to define.**

Hypothesis driven or exploratory studies

Before discussing how to effectively state the goals of non-hypothesis-driven functional genomics studies, it is worth reflecting on the common practice of framing many genomic studies as if they are hypothesis-driven. I believe that this tendency arises from the way scientific training has historically emphasized hypothesis-driven research as the cornerstone of rigorous science. With the rise of genomic studies, this tradition led to a strong preference, and at times, pressure, to describe studies as testing specific hypotheses, even when their goals are more exploratory in nature.

For example, a genomic study might be framed as testing the hypothesis that a particular treatment results in changes to gene expression in certain cells. While this framing aligns with traditional scientific language, it often pushes the concept of a hypothesis far beyond its intended meaning. In reality, such studies typically aim to address broader, exploratory questions, such as "What happens to these cells when we treat them in this way?" This is not a narrowly defined prediction but rather an effort to characterize the general outcomes of a specific intervention.

The reason these goals are more aligned with exploration than with hypothesis testing lies in the nature of what we are trying to uncover. Most likely, we lack specific expectations or clear ideas about which genes in these cells might be differentially expressed following the treatment. Instead, we are open to interpreting any pattern of differential expression as a potential outcome of the study. This makes the criteria for a productive study fundamentally different from traditional hypothesis-driven experiments. When we studied the impact of dietary protein levels on fish growth, for example, the only productive experiment is one that tests the specific hypothesis that fish fed a high-protein diet will grow larger.

In truly hypothesis-driven studies, we must test a specific prediction. Designing a controlled experiment to isolate the relationship between the intervention and the defined outcome is relatively straightforward, as the range of potential confounders is typically fairly narrow and well-defined. In contrast, framing exploratory functional genomics studies as hypothesis-driven is akin to a child pressing a button to see what happens, claiming to test the hypothesis that pressing the button will cause something, without having any idea what that something might be. We might consider the child's actions as hypothesis testing in a broad sense, but since anything could happen, ensuring that whatever occurs is truly the result of pressing the button and not due to some unrelated factor is an extremely challenging task. The range of potential confounders is virtually limitless.

Similarly, in functional genomics, we are typically not testing a narrowly defined hypothesis but rather exploring for any signal of interest, much like the child with the button. This exploratory approach demands a study design that prioritizes both facilitating discovery and eliminating confounders, ensuring that observed patterns can be reliably attributed to the treatment or condition being studied rather than to unrelated factors. Recognizing the exploratory nature of functional genomics studies can help us approach their design with greater clarity and purpose. By moving away from overly stretched notions of hypothesis testing, we can better align our methods with the true goals of the research, ensuring more robust and interpretable results.

Acknowledging this distinction is not a critique of exploratory research; it is a call for clarity and transparency about the true goals of these studies. By being precise about whether a study is hypothesis-driven or exploratory, we can improve study design, better address potential confounders, and communicate our findings and implications more effectively.

With that in mind, clearly articulating the goals of exploratory functional genomics studies can be challenging, and evidence from the literature suggests that these goals are often not stated with sufficient care. Unlike hypothesis-driven studies, it is difficult to provide general guidance for the effective design of exploratory studies because they vary widely in structure and objectives. **The goal of functional genomics studies is inherently tied to their design**. Unlike hypothesis-driven studies, which are guided by specific predictions, exploratory studies aim to uncover any pattern of interest. Consequently, the primary focus is on ensuring that observed patterns arise from the factors being studied rather than from unrelated or confounding influences. **In this context, the goal, scope, and design of a functional genomics study are essentially synonymous.**

Remember my comment regarding the confusion of the students when I asked for examples of the goals of functional genomics studies? This overlap between goals, scope, and design is what confused the students in my course and caused them to hesitate when answering what seemed like a straightforward question. As they considered the goals of a genomic study, the students realized they were implicitly beginning to design the study. Clearly defining the scope is essential for setting the boundaries of exploration and determining the parameters necessary to attribute observed patterns to the intended variables.

If I described a genomics study as designed to identify differentially expressed genes across human tissues, this casual statement *does not* provide enough intuition to convey the essence of the study. Instead, the goal and scope should be clearly defined, for example, as identifying genes whose expression levels differ between two specific tissue types, such as liver and kidney, using RNA sequencing to estimate gene expression levels from adult human donors. The goal, and the study design, should specify key parameters, including the population being studied (for example, healthy adult donors aged 30–50), the experimental approach for tissue collection (for example, paired tissues from the same donor), and the conditions under which the data are collected (for example, standardized protocols to minimize technical variability).

In addition, the study must account for potential confounders, such as age, sex, or batch effects in sample processing – we will discuss all of these properties in detail in the following chapters. By aligning the scope, goal, and design in this way, the study ensures that observed patterns in gene expression can be reliably attributed to biological differences between the tissues rather than to technical artifacts or confounding factors.

While the effective design of non-hypothesis-driven studies – and hence the goals - varies considerably, there is a crucial general warning that applies to all such studies: **one must never collect data for an exploratory study and then decide, after the data**

have been collected and analyzed, to revise the goal and use the data for testing specific hypotheses. This often occurs when the original goal of a study was simply to gather data rather than to gain specific insights or explore specific patterns. When hypotheses are formulated after examining the data, they are not chosen randomly. Instead, they are influenced by an initial review of the data, which may reveal patterns or correlations suggesting that these specific hypotheses are more likely to be supported.

This preliminary examination of the data involves an untold number of implicit tests, making proper statistical correction for multiple testing impossible. **Implicit multiple testing** can be a significant concern in genomics because readers cannot easily determine whether a published study was designed to address specific goals from the outset, or whether the data were first collected and explored, prompting the authors to formulate questions based on the observed patterns. The latter approach is particularly prone to producing misleading insights.

I should note that in functional genomics research, it is not uncommon to rely on publicly available datasets rather than collecting new data. This approach involves selecting existing datasets that are suitable for addressing specific research questions or testing well-defined hypotheses. While these datasets were often originally generated for exploratory studies or purposes unrelated to the new research questions, their use in a new context is entirely appropriate as long as the hypothesis is developed independently of the data. We will not discuss studies that involve the analysis of pre-existing data in this primer, as they do not involve designing a new study but instead utilize datasets that are already available. Researchers conducting such analyses should carefully consider the original study design used to generate the data.

A closer look at hypothesis testing and exploratory studies

Let's take another moment to further explain the problem of *ad hoc* hypothesis testing in exploratory studies. While this is a slight detour from the discussion of questions, goals, and study designs, it is a closely related topic and worth exploring in detail, as it may not be immediately intuitive. This is another example of how standard language can be imprecise, often leading to the wrong intuition. When we analyze RNA sequencing data, for example, it is common to say that we are testing whether genes are differentially expressed, but this isn't entirely accurate. In practice, hypothesis testing is an indirect process, a concept that may initially seem counterintuitive.

The original hypothesis is not tested directly. Instead, a null hypothesis is formulated to represent a pattern that contrasts with or opposes the original expectation. Data are then used to test the null hypothesis. If statistical analysis results in the rejection of the null hypothesis, this rejection lends indirect support to the alternative hypothesis, which reflects the expectation under the original hypothesis. This can be a bit confusing. An example will help clarify this: if we hypothesize that a certain treatment will result in specific differences in gene regulation between treated and untreated samples, we will use data to test the *null hypothesis* that there are no regulatory differences between the two groups. This null hypothesis is tested separately for each gene for which we have collected expression data.

Although this was not always the case, the need to account for multiple testing is now generally well appreciated in genomics. We understand that if we perform 100 independent tests of a null hypothesis, whether by testing 100 different hypotheses or independently testing a single hypothesis 100 times, we expect to reject the null hypothesis, and thus provide apparent support for the alternative hypothesis, approximately five times on average when using a *P value* threshold of 0.05. Because of this, we adjust the statistical cutoff used to reject the null hypothesis to account for multiple testing. By doing so, we minimize the probability of false positives and improve the reliability of our conclusions.

Implicit multiple testing occurs when hypotheses are selected for testing after the data have been collected and examined, rather than being defined beforehand. Once the data are available, it may become clear that certain hypotheses are unlikely to be supported, reducing the motivation to formally test them. Conversely, other hypotheses, including some that might not have been considered initially, may appear more plausible based on patterns observed in the data, prompting a desire to test them explicitly. This preliminary examination of the data increases the likelihood that we only choose to test hypotheses that are more likely to be supported.

This is not an appropriate way to test hypotheses. The standard statistical methods we rely on assume that the data are generated independently from the process by which the hypotheses are formulated. These statistical methods are calibrated under this assumption, ensuring that when applied to completely null data (for example, data in which there truly are no differentially expressed genes), we expect to reject the null hypothesis 5% of the time by chance, with a *P value* of 0.05. In other words, in 5% of cases, the dataset will appear unusual and not representative of the broader set of all possible datasets we could have collected by chance alone. For a well-calibrated test, this probability - one in twenty, or 5% - is considered an acceptable margin of error.

However, if we test 100 hypotheses after observing the data, many would appear to be supported because their formulation was influenced by the data itself. Implicitly, we have already tested numerous hypotheses by discarding those we anticipated would not be supported. By failing to document these implicit tests, we increase the likelihood of false positives, as we cannot accurately account for the extent of multiple testing or properly control for the possibility that some observed patterns are merely due to chance. This conduct can lead to an inability to reproduce published observations.

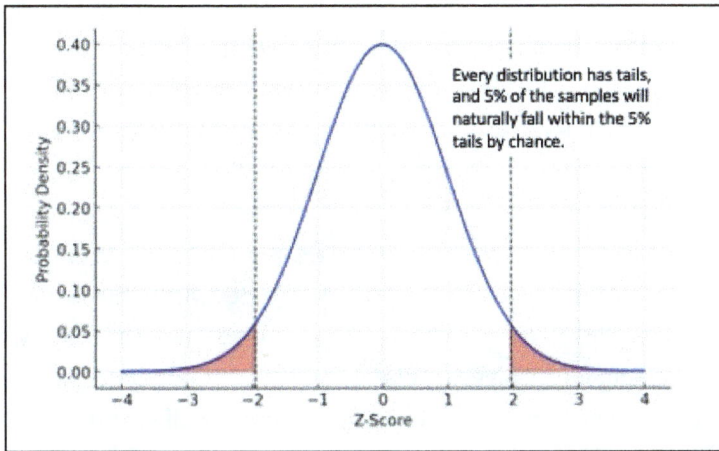

To avoid this issue, data collected for non-hypothesis-driven studies should not be directly used to test hypotheses formulated *post hoc*. Instead, such studies should serve as a basis for generating hypotheses, which can then be tested in hypothesis-driven studies using newly collected, independent datasets.

Functional genomics and the study of biological mechanisms

Since we are discussing the task of defining the goals of functional genomics studies, it is worth taking a moment to address the topic of mechanistic insights. This is a contentious issue for many. Scientists who primarily use standard molecular biology techniques often criticize functional genomics for failing to generate actionable mechanistic insights. In response, functional genomics scientists defend their field by emphasizing the mechanistic insights that can only be revealed through the large-scale data and exploratory approaches inherent to their methods.

This debate is further complicated by the inherently subjective and evolving definition of novel mechanistic insight. The definition of novel mechanism has evolved with technological advances and the new questions they allow us to explore. In the early days of genomics, when tools were limited to analyzing DNA sequences, mechanistic questions focused on the relationship between genes and function. Studies that resulted in the annotation of broad regions of the genome as functional, were often considered lacking in novel mechanistic insights until specific genes within those regions were linked to specific phenotypes or functions. Ironically, this was frequently achieved without functional genomics data, meaning no actual functional mechanisms were studied at the time.

As genomics progressed and technologies for gene expression profiling became widely available, even these data were often said to be insufficient for providing new mechanistic insights. By that time, the definition of biological mechanism had evolved further to emphasize the identification of regulatory elements, protein interactions, and signaling networks. With each technological advance, the standard for what qualifies as novel mechanistic insight has continued to rise.

Today, functional genomics studies are often regarded as lacking novel mechanistic insight unless they provide an understanding of specific molecular events, such as how particular transcription factors or protein complexes drive specific gene expression changes and, ultimately, specific cellular outcomes.

This is why I typically avoid discussing mechanisms when defining the goals of a functional genomics study. Frankly, I'm not sure I know what "mechanism" means exactly, and as a result, it's not a useful tool for setting clear objectives. Instead, I prefer to describe the goals of a study, or the expected results, in terms of correlation and causation.

Functional genomics studies are exceptionally useful for uncovering new patterns in the data, such as co-expression of genes, shared pathway activities, or epigenetic changes linked to a phenotype. However, these patterns are typically correlational and do not establish causation. Even experiments involving manipulations, such as gene knockouts or treatments, often produce data that remain fundamentally correlational. For example, knocking out *Gene A* and observing changes in the expression level of *Gene B* might imply a regulatory relationship, but without further evidence, such as direct binding studies or functional pathway analyses, such conclusions remain quite speculative.

Functional genomics truly excels as a hypothesis-generating tool, providing large-scale, high-dimensional datasets that can help identify potential causative relationships for further testing. Returning to the child and the button example, imagine that when the child first pressed the button, a nearby phone rang. This was the child's exploratory, correlation-based study. Based on this observation, the child might form a hypothesis that pressing the button causes the phone to ring. The child can now conduct a new, well-replicated hypothesis-driven experiment to test this specific hypothesis, by repeatedly pressing the button to see if the phone rings each time.

In genomics, a real example comes from studies that uncovered previously unappreciated correlations between different histone modifications and gene expression levels. These findings led to putative mechanistic insights and the generation of specific causative hypotheses about the pathways and regulatory interactions underlying these patterns. Researchers tested and validated these hypotheses through targeted experiments, such as perturbation studies and protein-DNA binding assays, advancing our understanding of causative relationships and the specific roles of histone modifications in gene regulation.

The understanding of what functional genomics studies can and cannot achieve is essential when defining the goals of a study. We can be specific in designing a study aimed at causative inference or mechanistic insight while recognizing that such insights will initially be based on correlations and must be validated further. By using precise language to articulate the goals of the study, the expected insights, and their limitations, we can maximize the information extracted and design the most effective study to achieve those insights.

Problems arise when we fail to define specific goals and instead decide which questions to ask only after examining the data, or when we overlook the fact that our findings are correlation-based and mistakenly present them as causative mechanisms. Clear goals and an awareness of the limitations are key for designing effective studies and producing meaningful and reliable results.

Designing with purpose (I): what to expect from genomic technologies?

After defining the goals of our study, we must carefully consider the methodology for data collection. It is essential to understand both the type and the significance of the data we plan to collect. Often, the standard terminology used to describe data can be misleading (this is a reoccurring theme, I realize). For example, terms like "studies of gene expression" or "measuring gene expression" frequently arise in discussions of studies that involve the collection of RNA sequencing data. These phrases might suggest that RNA sequencing can be used to measure the absolute levels of gene expression. However, in practice, it is highly challenging to measure absolute gene expression levels using RNA sequencing; in fact, I am not aware of any studies that successfully do so. Instead, RNA sequencing is typically used to measure relative gene expression levels.

The discussion of precise terminology and definitions in the context of data collection often surprises the students in my course. For instance, when I ask, "What do we aim to measure when we collect bulk RNA sequencing data?" students are frequently confused when I do not accept the answer "gene expression." Instead, I am looking for the more precise response: "relative gene expression levels." When we first discuss study design, many students fail to grasp the importance of what may seem like subtle or negligible distinctions between imprecise and precise answers. However, precise definitions, and more importantly, an understanding of the specific properties and limitations of each technology, are essential for effective study design.

The term "absolute expression level" refers to specific, quantifiable measurements, such as the number of RNA molecules per cell or the concentration of a specific RNA type, expressed in defined units. In contrast, "relative expression levels" represent the proportion of RNA from a particular gene relative to the total RNA expressed from all genes in a sample. While the distinction between absolute and relative expression levels may seem subtle, it is critically important.

For those who remember the early days of DNA microarray studies, you might recall the widespread misunderstanding of probe effects (the specific oligonucleotide sequences on a microarray designed to hybridize with complementary DNA or RNA target). I vividly remember colleagues asking for help with experiments aimed at determining which genes were expressed in a specific tissue or cell type, and their confusion when I explained that microarrays couldn't be used to effectively answer that question. How could that be? Microarrays were developed and widely used precisely to compare gene expression *between samples*; so why couldn't they be used to answer what seemed like the simpler question of whether a specific gene is expressed in a *given sample*?

The answer is related to challenges involving hybridization kinetics and probe design. While the intensity of hybridization of specific RNA molecules (transcripts from particular genes) to their corresponding probes was correlated with the abundance of those RNA molecules, it was more strongly influenced by the inherent properties of the probes themselves, such as their composition and hybridization kinetics. This influence, referred to as the probe effect, introduced significant challenges. Because the probe properties had such a strong impact on hybridization kinetics, it was unreliable to interpret hybridization intensity from any given probe as a direct measure of absolute gene expression. In fact, the background signal on some probes could exceed the actual hybridization signal on other probes. In other words, probes for some genes that were not expressed at all could display higher intensity due to background hybridization than probes for genes expressed at relatively high levels. Consequently, using microarrays to determine whether a gene was expressed based solely on hybridization intensity proved to be problematic.

Microarrays were well-suited for comparing relative hybridization intensities across samples using the same probe on different arrays. By doing so, the probe effect could be ignored, as it remained constant for a given probe across samples and arrays. This allowed researchers to determine relative differences in gene expression between conditions. However, it also meant that the only way to conclude that a gene was truly expressed in a sample was by comparing its expression profile to another sample, observing differences in relative expression. In other words, microarrays could be used to reliably identify whether a gene was more highly expressed in one sample compared to another, but they could not be used to confirm whether a gene was expressed at all without this comparative context.

Like microarrays, RNA sequencing can be used to estimate relative expression levels but not absolute ones. While read counts in RNA sequencing are not susceptible to probe effects, they represent a finite resource distributed across all expressed genes within a sample. The read count for any specific gene must be interpreted in the context of the total read count for the sample, as it reflects a relative measure rather than an absolute quantity.

When analyzing RNA sequencing data, two genes, for example, might show the same relative expression level across different conditions, yet their absolute expression levels could differ significantly if the baseline expression levels are not comparable. This reliance on relative data underscores the need for rigorous procedures to standardize and normalize samples, ensuring meaningful comparisons. While standardization and normalization are essential, normalization requires the assumption that the majority of genes are not differentially expressed between conditions, an assumption that may not always hold true. We will explore normalization methods and their associated assumptions in greater detail later in this primer. Recognizing the importance of standardization and normalization shapes both the design and analytical strategies of genomic studies.

Understanding the limitations of relying on relative gene expression levels is essential for designing RNA sequencing studies and interpreting their results. Relative expression data enables comparisons of the rank order of genes within and across samples, but it does not provide estimates of the actual counts or concentrations of RNA transcripts. This limitation implies that conclusions about the magnitude of expression changes or their biological significance are based on assumptions that may not be valid across different experimental conditions or sample types; in the following chapters, we will discuss examples of these exceptions and explore some of the consequences of misunderstanding their implications.

Though this primer focuses largely on examples involving gene expression data, the distinction between relative and absolute quantification is equally critical across other functional genomics data types. In ChIP-seq (Chromatin Immunoprecipitation sequencing) studies, for example, interpreting chromatin enrichment requires an understanding that the data represent relative measurements. These measurements reflect the binding affinity of DNA-associated proteins or the presence of specific histone modifications, typically assessed in comparison to an input control sample (a sample from which cross-linked DNA is extracted without immunoprecipitation). While ChIP-seq analyses provide valuable insights into enrichment patterns, they do not convey the absolute number of binding events or the precise concentration of the factor at a specific site.

The case of the cell culture, the treatment, and the independent biological sample

> **What I assume you already know:** In this chapter, I assume the reader has prior knowledge of standard laboratory techniques and workflows related to cell culture and RNA sequencing. The discussion builds on the understanding that immortalized cell lines serve as controlled model systems. Basic experimental practices such as sub-culturing cells, maintaining optimal conditions in incubators, and using nutrient-rich media are expected to be familiar. Similarly, the concept of applying treatments to cell cultures at specified concentrations, alongside untreated controls for comparison, is expected to be understood. RNA extraction and sequencing are central to this example, and I assume familiarity with the basic concepts of isolating high-quality RNA, preparing libraries, and performing bulk RNA sequencing.

The next question I ask the students in my course requires some setup. So, let's begin designing our first functional genomics study. We need to start with a specific goal, which, in this case, we define as studying the regulatory impact of treatment X on gene expression in a cell line. However, as we have discussed earlier, this goal is not specific enough. Let's refine it further:

We aim to design a study to characterize gene expression differences in a single human immortalized B cell line (we are <u>not</u> interested in a general response of all B cell lines across donors, just this specific line) following treatment with compound X in a given specific concentration for a given treatment time. We will collect bulk RNA sequencing data to obtain expression measurements.

This revised goal should be specific enough for our purpose. The key question I want to explore in this section is: ***What*** *constitutes an independent biological sample?* Often, standard language falls short in this context. For example, one might answer my question by stating that we will compare treated and untreated cultures of the B cell line under investigation. But what exactly do we mean by "cultures"? To design an effective study, we need to clearly define what constitutes an independent biological sample in this scenario.

In scientific research, an "independent" sample is one whose data do not directly dictate or determine the data of another sample. Samples treated as independent within an experimental design may still exhibit correlations or shared patterns arising from factors like a common origin or similar environmental conditions. For example, in an experiment with a single cell line, cultures in different flasks may be considered independent because each flask is handled separately, such that data collected from cells cultured in one flask do not influence the data from another flask. However, these samples share a common biological origin, and their independence is contextual, relevant within the scope of that shared origin. Statistical independence is not an absolute property, but a construct defined relative to the objectives of the study.

In a different scenario involving cell lines from multiple individuals, cell cultures established from different individual donors would be considered independent, as they originate from distinct biological sources. The variation observed among donor-derived cell cultures reflects both biological differences and technical variations introduced during the cell culturing process, offering a broader context of independence. (If you are unfamiliar with the use of the terms "technical" and "biological" in this context, don't worry; we will discuss and explain these concepts in greater detail later in the book.) Understanding and defining the appropriate level of independence is essential for accurate data interpretation. It ensures that the variability captured in the study reflects true biological differences rather than systematic similarities that could bias the results.

In the context of an experiment using a single cell line, independent samples refer to separate sub-cultures derived from the same cell line. Let us now consider the size of the sub-cultures as these are going to be our independent experimental units. Given our aim to collect bulk RNA sequencing data, it is feasible to obtain high-quality results from as few as tens or hundreds of cells. But **what would be the optimal size of the cultures for our study?** This question is not as straightforward as it might seem, and the answer depends on several factors related to study design and biological variability.

Before you read further, take a moment to reflect on this question and try to formulate your own answer. What considerations might influence your choice of culture size?

Welcome back. Should we consider cultures of just hundreds of cells as the independent biological unit, or are larger cultures necessary? At the extremes, practical considerations come into play. While high-quality bulk RNA sequencing libraries can be generated from just few hundred cells, culturing and processing such a small number of cells is challenging in most settings. Without a specific reason, the effort to develop experimental infrastructure for such tiny cultures may not be worth the investment. At the other extreme, scaling up to maintain large cultures of billions of cells is expensive and resource intensive. Unless a specific need justifies the cost, there is little to gain from scaling up to this level. Determining the optimal culture size involves balancing practicality with the study's goals, ensuring that the chosen approach aligns with both experimental needs and available resources.

There is another reason to avoid using small cultures consisting of just a few dozen, hundreds, or even a few thousand cells. Since our goal is to study the bulk (as opposed to single cell) gene expression response of this specific cell line to treatment with compound X, we aim to capture an unbiased average (bulk) response. Small cultures may not accurately reflect the overall response of the cell line because they could, by chance, contain a disproportionate number of stressed cells, non-responsive cells, or cells with atypical responses. As a result, cultures with very small numbers of cells are likely to show substantial variability in their bulk gene expression data, making it challenging to precisely capture the cell line's overall regulatory response following the treatment. If small cultures are used as independent biological samples, the variability between samples would be high, requiring a much larger number of replicates to reliably assess the regulatory impact of treatment with compound X.

How large should the cultures be to minimize biological variance? Our goal is to strike a balance between efficiency and precision, avoiding unnecessary use of resources by not automatically defaulting to immensely large cultures if smaller ones can provide comparable results. How do we make this decision? While this question is specific to the example at hand, it represents a broader dilemma that often arises when defining the properties of study designs and determining what constitutes an independent biological sample. Take a moment to reflect on this question and consider how you might approach making this decision, both in this scenario and in similar cases.

Welcome back. To provide an informed answer to this question, one might need to conduct a pilot study. This approach is one I frequently rely on, not only as a theoretical solution, but also as a practical tool in my own research. While existing literature and our colleagues can often provide valuable guidance, there are instances where the specific information we need is unavailable, requiring us to seek answers through our own pilot studies.

In this case, the goal of a pilot study would be to identify the point of diminishing returns, determining the culture size at which further increases no longer yield cost-effective benefits. This could happen because increasing the size of the cultures further may not significantly reduce the variance observed between them, or because, beyond a certain point, it becomes more practical and cost-effective to establish additional replicate cultures (thereby improving variance estimation with more replicates) rather than continuing to scale up individual cultures.

In this specific cell culture example, however, we can draw on previously published studies and the collective experience of other scientists, making a dedicated pilot study unnecessary. It is well-established that bulk RNA sequencing data from sub-cultures of 100,000 cells or more, sampled from a larger culture, yield nearly identical estimates of gene expression levels. As a result, using cultures larger than 100,000 cells in bulk gene expression studies does not significantly reduce the variance between cultures and rarely results in detectable improvements in the data.

A single factor study design requires more than a single sample

Now that we understand the concept of independent samples and have determined the optimal culture size, let's focus on the specific experimental procedures for obtaining the culture and designing the treatment experiment. One of the first and most important considerations is the use of replicates. We previously mentioned replicates in the context of the expected differences between sub-cultures when discussing the optimal culture size. But why are replicates essential in any study design?

To demonstrate this, consider a scenario where we observe based on RNA sequencing data, that the relative expression level of a specific gene is X% higher in a **single** treated sub-culture compared to a **single** untreated (control) sub-culture of the same cell line. What does this mean? If X is small, say 10%, one might conclude that the relative expression levels of this gene in the treated and control cultures is quite similar. Conversely, if X is large, say 300%, one might assume that there is a meaningful difference in gene expression levels between the cultures. Does this inference seem correct to you? Take a moment to think about this.

Given the information provided about the relative expression levels of the gene in a single treated and a single untreated culture, both inferences - that a 10% difference indicates similar expression and that a 300% difference signifies a meaningful difference in expression - are **incorrect**. In either case, we lack sufficient information to determine whether the observed difference is meaningful or simply due to random variation.

Since our plan is to collect bulk gene expression data, conducting an experiment with just a **single** treated and a **single** untreated culture would yield only one gene expression measurement per condition. Without replicates, we would lack a benchmark to determine the expected variability between measurements, making it impossible to meaningfully assess any observed differences or similarities.

It is necessary to account for the natural variability between independent replicates of similar cultures, whether untreated or treated, to contextualize and interpret the results. This can only be accomplished through replicate measurements. If the variability between replicates of the same type of culture (untreated or treated) is as large as the gene expression differences observed between treated and untreated samples, we cannot confidently attribute the differences to the treatment itself; they might simply reflect the expected variation in gene expression between any cultures, regardless of treatment.

Moreover, natural variability in expression is often gene specific. Some genes exhibit highly consistent expression across different cultures, such that even a small treatment effect of 10% shift in gene expression, might exceed the range of natural variation and be biologically meaningful. Conversely, some genes show substantial natural variability, such that a measured treatment effect as large as 300% might not differ significantly from the variation typically observed across cultures, regardless of the treatment status. Without replicates, we have no way to determine whether the observed treatment effects are biologically meaningful or merely reflect the inherent natural variability in gene expression.

To determine the necessary number of replicates, we typically perform a power analysis. Although power analysis is an integral part of study design, I will not explore it in detail in this primer. Designing effective studies requires an understanding of both study design principles and power analysis. However, because power analysis often relies on simulations based on the study design itself, I have chosen to focus on study design in this text. I assume that readers aiming to design their own studies will seek additional resources to learn about power analysis.

For our current discussion, let us assume that a power analysis has indicated that three replicates each of treated and untreated cultures are needed. This information is sufficient for our purpose because we are studying the impact of treatment with compound X on gene expression in a specific cell line. Since our goal does not extend to examining variability across different individuals, there is no need for biological replicates derived from multiple donors.

It is also worth noting that one of the reasons we choose cultures of 100,000 cells is to minimize the natural variability between replicate cultures of the same kind. With this culture size, the variability between replicates of the same type - treated or untreated - is expected to be low. This reduced variability helps increase the statistical power of the study, allowing meaningful conclusions to be drawn with a relatively small number of replicates.

How should we design the replicated experiment?

Consider the three scenarios illustrated in the figure on page 45. The X-axis shows the timeline of experimental procedures, with steps occurring later in the experiment positioned further to the right. Replicates are indicated by multiple images; for example, in the first scenario (top panel), we extract RNA from the original large untreated culture in three replicates. At the same time, we also establish from the original culture three smaller replicate sub-cultures and treat them with compound X for a given time, then we extract RNA from each sub-culture.

In all three scenarios, RNA sequencing is performed on samples from three treated and three untreated cell populations. The resources required are comparable across all three scenarios. The key differences between the three scenarios lie in the timing of various processing steps. **Our objective is to evaluate the impact of these procedural choices and select the most effective study design given our experimental goals**.

To facilitate this exercise, I have simplified one aspect: in the power analysis previously mentioned, which determined the need for three replicates each from treated and untreated samples, a decision about which scenario to apply would have already been made. (We have to specify the study design when we estimate the power of the study…) However, for the purpose of this exercise, let's overlook this complexity. We will fix the size of the study to six samples in total and focus on determining which setup yields the most effective design. In essence, we are asking: **Given an identical allocation of resources, which study design will maximize the power to** detect differences between treated and untreated cultures and ultimately be the most effective?

Take a moment to consider the figure on the next page, evaluate the differences between the study design scenarios, and choose the most effective design before you read further.

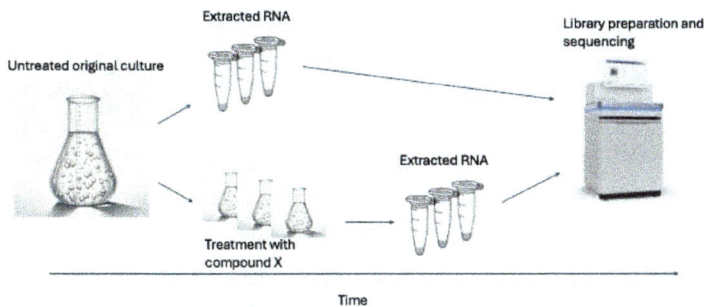

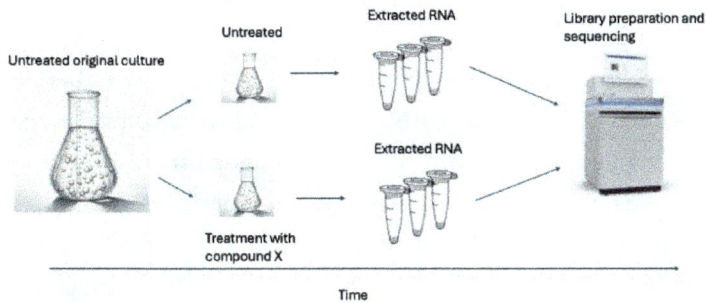

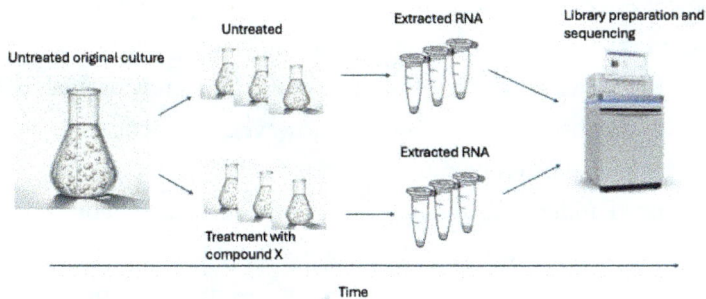

Let's take a closer look at the first scenario (top panel). In this design, RNA is extracted in three replicates from untreated samples taken directly from the original culture, without additional sub-culturing. These RNA samples are extracted and processed immediately. Meanwhile, three new replicate sub-cultures are established from the original culture and treated with compound X for a specified duration. RNA from the three treated sub-cultures is extracted at a later time point.

This design is flawed because it introduces two major confounders related to the primary parameter of interest, the treatment effect. First, the duration of culturing is confounded with the treatment. RNA is extracted from the untreated cells immediately, while the treated cells undergo an additional culturing period. As a result, any observed differences in gene expression levels could be attributed to the extended culturing time rather than the treatment itself. Second, RNA extraction is confounded with the treatment, as it is performed in two separate batches, one for the untreated samples and another for the treated samples. This creates the potential for batch effects, where differences in gene expression may arise from inconsistencies in the RNA extraction procedure rather than the treatment. The critical flaw in this first study design is that these confounders are fully aligned with the treatment effect, making it impossible to disentangle the true impact of the treatment on gene regulation.

In this example, the confounders in the first study design are relatively straightforward to identify. However, in practice, confounders can often be quite subtle and difficult to detect. It is important to understand that any systematic difference between samples, whether related to timing, handling, or processing conditions, can act as a confounder. Even seemingly minor variations have the potential to create confounders that might mistakenly be interpreted as true treatment effects. To ensure that observed differences in gene expression are genuinely attributable to the treatment, careful consideration and meticulous planning are required. Anticipating potential confounders and structuring experiments to avoid or minimize them is essential for producing reliable and meaningful results.

Consider the second and third scenarios (middle and bottom panels). In both, the culturing time is identical for treated and untreated cells, and RNA extraction for all samples occurs simultaneously in a single batch. As a result, neither scenario introduces a technical confounder related to the treatment effect. The key difference between the second and third designs lies in how the three treated and untreated replicates are established. **In the second scenario** (middle panel), we simultaneously culture one treated and one untreated sample (sub-culture) for the same duration. After the treatment period, three RNA extraction replicates are processed from both treated and untreated cultures. **In the third scenario** (bottom panel), three replicate sub-cultures are established before the treatment period begins, and each sub-culture is processed separately for RNA extraction after the same treatment duration. Although the total number of samples processed in all scenarios (including the first) is the same, the difference in the nature of replication – treatment or RNA extraction - significantly affects the utility of the study.

To appreciate the difference between the second and third scenarios, it is helpful to recall the study's stated goal: to assess the treatment's impact on the cell line. In the second scenario, a single treated sub-culture and a single untreated sub-culture are used, with three RNA extractions performed for each. This design allows for precise estimation of variance due to sample processing, specifically RNA extraction and sequencing, because we replicated it three times. Additionally, by measuring gene expression from each culture three times, we obtain precise estimates of gene expression differences between the single treated and untreated cultures. However, the observed regulatory differences may be specific to these individual sub-cultures, arising by chance, and may not reflect the overall impact of the treatment on the cell line as a whole. While this precise measurement of variance associated with sample processing is valuable, it reflects only the variance from RNA extraction and sequencing, rather than providing generalizable insights into the treatment effect on the entire cell line.

In contrast, the third scenario involves using three independent sub-culture replicates for both the treated and untreated conditions, providing a basis for generalizing the findings to the entire cell line. In this design, gene expression is measured only once for each sub-culture, so the variance associated with RNA extraction and sequencing is combined with the biological differences between sub-cultures. As a result, the variance estimates are less precise than those in the second scenario. However, this approach better supports the study's primary objective, which is not to separate variance components but to gain a general understanding of the treatment's overall impact on the cell line.

By considering these different scenarios we learned that designing a treatment experiment using cell cultures requires a thoughtful balance between precision and generalizability. The first scenario highlights the severe consequences of complete confounders, which can invalidate an experiment and render its results unusable. Without addressing confounders, any observed effects cannot be reliably attributed to the treatment, undermining the study's validity. The second scenario demonstrates the value of precisely measuring variance associated with the technical aspects of sample processing, but it falls short in providing broader insights into the treatment's general impact on the cell line. The third scenario results in the confounding of different sources of variance, which is acceptable as they are not the focus of the study.

The third study design offers a more robust framework for generalizing findings about the response of the cell line to the treatment. Ultimately, the choice of experimental design should align with the study's primary goals, ensuring that it minimizes confounders while achieving the desired outcomes.

Other potential confounders to consider

Up to this point, the discussion has centered on the structural aspects of study design, particularly the timeline and the nature of replicates. However, effective experimental design encompasses more than just these elements. Critical details related to the actual implementation of the study also play a vital role. Additionally, various biological and environmental parameters must be considered, as they can introduce significant sources of variation or confounding factors. Before addressing the practical details of study implementation, we should consider the concepts of bias and variance, as they are central to the decision-making process in experimental design.

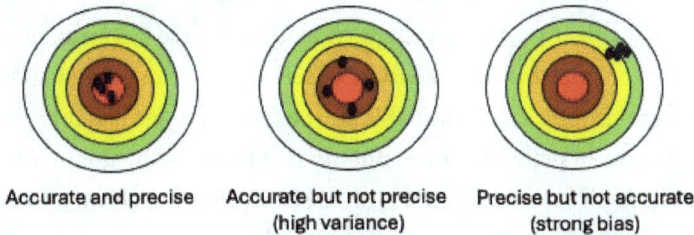

| Accurate and precise | Accurate but not precise (high variance) | Precise but not accurate (strong bias) |

Ideally, both bias and variance should be minimized to ensure results that are accurate and precise. In practice, however, trade-offs are often necessary. **To prevent the introduction of confounders, which can lead to bias, it may be necessary to tolerate a certain level of increased variance.**

While confounders can invalidate an experiment by making it impossible to attribute any observed effects to the treatment, increased variance merely reduces measurement precision. This reduction in precision lowers the statistical power of the study, often requiring a larger sample size to detect a true effect. Successfully understanding and managing this balance is essential for designing robust and reliable experiments.

In addition to the structural aspects of study design, the practical execution of the experiment plays a significant role in ensuring that the results are reliable and reproducible. One important consideration is determining who will conduct the experiment and the specific timing of each step. In our example, we have designed a study that can be completed within a single day, which helps simplify many aspects of execution and minimizes potential complications. However, even in this relatively straightforward scenario, several factors must be carefully controlled to avoid introducing confounders that could compromise the results. We will address the additional complexities of multi-day experiments in later sections, the focus here is on the considerations relevant to this single-day setup.

Whenever possible, a single individual should perform all the work to ensure that any variation observed between treated and untreated samples is attributable to the treatment itself rather than differences in technique or handling. If multiple individuals are needed, their tasks must be evenly distributed across both treated and untreated samples – we call this *'balancing'* - to prevent systematic differences. Assigning all treated samples to one person and all untreated samples to another, risks introducing a confounder, as observed differences may reflect individual techniques rather than treatment effects. While balancing, namely evenly distributing tasks among individuals increases variance, it avoids introducing bias. If necessary, the sample size can be increased to compensate for the reduced power.

Consistency in the materials and reagents used is also important to avoid confounding effects. All chemicals, kits, culture medium, and other resources should be harmonized across treated and untreated samples to prevent batch effects. The ideal approach is to use the same batch of reagents for all samples. If this is not feasible, randomizing or balancing the use of different batches across all samples is an acceptable alternative. While this may result in increased variance, it avoids introducing bias - a principle that echoes throughout this primer and will undoubtedly resurface.

In addition to the practical aspects of performing the experiment, environmental factors that could introduce variability or confounders must also be carefully considered. One critical environmental factor is ensuring that cells are cultured under consistent conditions. Parameters such as temperature, CO_2 levels, and humidity must be identical for both treated and untreated samples. Ideally, all cultures should be housed in the same incubator to maintain uniform conditions. If multiple incubators are required, it is important - at the risk of sounding like a broken record - to avoid separating treated and untreated samples by incubator, in order to avoid a confounding factor.

Turning to biological aspects, one important factor to consider is the cell cycle. Most cell types in culture naturally exist in a mixture of different phases - G1, S, G2, and M phases - each characterized by distinct transcriptional states and gene expression profiles. This raises two potential concerns. First, different cells in different phases of the cell cycle may respond differently to the treatment, as the transcriptional activity of a cell in G1 could be quite different from one in the S phase. Second, there is a risk that treated and untreated cultures may have different distributions of cells across the stages of the cell cycle. This discrepancy could result in differences in gene expression that are not caused by the treatment itself but rather by the varying proportions of cells in different phases of the cell cycle.

What can be done to address this issue? One approach is to use techniques that synchronize cell cultures, ensuring that all cells in a culture are at the same phase of the cell cycle at a given time. Another option is to arrest cells at a specific phase, halting their progression through the cycle. For instance, chemical treatments can be applied to stop cells in the G1 phase, preventing their transition into the S phase where DNA replication occurs. These methods aim to create more uniform conditions across the experiment, potentially reducing variability introduced by differences in cell cycle phase.

Before we proceed, take a moment to consider: The synchronization and arrest methods we mentioned can achieve a more uniform composition of cell cultures. Would it be better to synchronize the cultures, arrest them at a specific phase, or leave them as they are, maintaining their natural cell cycle distribution?

The correct answer comes back to the fundamental principle we always rely on, namely, the goal of the study. In this case, the goal is to understand the natural response of the cell line to the treatment. If we choose to arrest the cells at a specific phase, we artificially freeze them in a non-physiological state, and any response to the treatment may be specific to that artificial state. This would not represent how the cells normally behave under natural conditions. Similarly, if we synchronize the cultures so that all the cells are in the same phase, we might think we're just allowing them to resume a natural state after synchronization. However, this would only reveal how cells at a particular state respond to the treatment, which may not provide a generalizable answer to how the cell line, as a whole, responds.

More critically, both approaches, synchronization and arrest, must be flawless to avoid introducing confounders, but in reality, they rarely achieve this level of perfection. Small differences in the effectiveness of synchronization or arrest between treated and untreated samples can result in confounding effects. For example, if the synchronization process happens to work slightly better in the untreated samples - and I'm sure you can see where this is going - this discrepancy could falsely indicate a treatment effect when none actually exists.

When working with cultures of approximately 100,000 cells in their natural, mixed states, it is unlikely that the cell cycle distribution will differ significantly between any two cultures, including between treated and untreated samples. Even if minor, random deviations occur, they are unlikely to consistently correlate with the treatment in a well-replicated experiment, so the risk of confounding remains minimal. Paradoxically, attempts to impose uniformity through synchronization or arrest might actually introduce confounders in this case. Small, unintended differences in the effectiveness of synchronization or arrest between groups can significantly impact the results, potentially introducing variability that falsely appears to be associated with the treatment.

Considering all these factors, the optimal design in this case may ultimately involve synchronization, followed by an experimental design that evaluates treatment effects within each phase of the cell cycle. However, to do this, we will have to compare treated and untreated cells at each cell cycle phase, separately, using sufficient replicates that are independently synchronized to reduce the risk of confounding due to imperfect synchronization. This approach would be extremely resource-intensive, as it requires enough replicates to mitigate any inconsistencies in synchronization. This approach effectively amounts to conducting an independent well-replicated experiment for each cell cycle phase.

The compromise, and likely the most practical option, is to leave the cultures in their natural state, acknowledging that different phases of the cell cycle may respond differently to the treatment. In this scenario, the observed bulk response represents an average across all cell cycle phases, potentially resulting in higher variance compared to a synchronized culture. However, this added variance is a relatively small price to pay for a much simpler and more efficient experiment that avoids the risk of introducing a confounder due to imperfect synchronization or arrest.

What have we learned from this example of study design?

The first example of study design in the primer focused on simple study of the treatment effect in a single cell line. This simple example illustrates several foundational principles of effective study design in functional genomics. By focusing on the nuanced challenges and the decisions involved, it provided us with insights that transcend the specifics of this experiment and apply broadly to many study types.

Key lessons include:

- **Defining independent biological samples**: We discussed the need to precisely identify what constitutes an independent sample in the context of the study. Whether optimizing culture size or controlling biological variance, thoughtful decision-making is essential to balance experimental rigor with practicality. In cases of uncertainty, pilot studies offer a reliable path to informed design.

- **Mitigating confounders through careful planning**: We emphasized that even subtle procedural differences, such as discrepancies in culturing time, RNA extraction, or sample handling, can confound results. Careful attention to these details ensures that observed differences reflect the treatment effect, not unintended biases.

- **Navigating the trade-off between precision and generality**: The comparison of different experimental designs highlights the importance of aligning the approach with the goals of the study. While precise measurement of technical variance has value, the broader goal of generalizing findings across biological replicates was more effectively achieved with a design that prioritized biological variability over technical exactness.

- **Aligning practical design with research goals**: The deliberate decision to observe natural cell cycle variability, rather than impose uniformity through synchronization or arrest, underscores the importance of tailoring experimental conditions to the study's objectives. This approach ensured that the findings reflected an authentic cellular response to the treatment.

Designing with purpose (II): biological vs. technical replicates

The first experiment we designed was relatively simple, involving a single treatment applied to one cell line. The scope of the experiment was confined to this specific cell line, with no intention of generalizing the treatment effect to other cell lines or individuals. This design allowed us to explore key concepts such as independent samples, experimental workflows, and methods for controlling simple confounders. As we move on to discuss more complex study designs that include multiple individuals and different conditions, it becomes important to carefully consider the role of replicates across different levels of the study.

It is helpful to clarify the terms "biological replicates" and "technical replicates." While these terms are widely used, their exact meanings can vary depending on the context of the study. Generally, technical replicates are designed to account for variability introduced during sample processing or experimental procedures, whereas biological replicates reflect natural biological variability. However, the distinction between the two is somewhat fluid and depends heavily on the experimental design and the research questions being addressed. In practice, labeling a specific level of replication as technical or biological is less important than designing the study thoughtfully.

In the study design example involving a single cell line, the independent sub-cultures are considered biological replicates, representing distinct samples derived from the same cell line. These replicates were used to capture the biological variability that naturally exists between independent sub-cultures. In contrast, when multiple RNA extractions were performed from a single culture, the resulting samples are considered technical replicates, aimed at assessing variability introduced by the experimental procedures and sample processing.

In some versions of the experimental design, we deliberately combined different aspects of technical variation. For instance, instead of isolating the effects of RNA extraction and sequencing library preparation, we treated them as a single source of technical variability. This approach was intentionally selected because our goal was not to estimate the variability of each individual step. Instead, our focus was on accounting for the overall technical variability introduced throughout the experiment.

In one version of the design, we intentionally chose to confound both biological and technical variation. We established biological replicates through independent sub-cultures and performed a single RNA extraction from each sub-culture. In this case, the observed variation in gene expression between sub-cultures represented a combination of both biological and technical variability. In this setup, the two sources of variation were not separated. Although the sub-cultures served as biological replicates, they were used to capture the overall variance in the experiment, encompassing both biological and technical components.

By combining biological and technical variation, we sacrificed some precision in each replicate, as the added technical variability increased the range of measurements. However, foregoing technical replicates allowed us to allocate more resources toward biological replicates. This decision was guided by prior experience, which indicated that in this system, biological variance between sub-cultures far exceeded the technical variance introduced by sample processing. Given the objectives of our study, prioritizing biological replication over reducing technical variation was expected to yield the most meaningful results.

As we scale up to more complex experiments, particularly those involving multiple individuals, the distinction between biological and technical replicates becomes progressively less clear. What we annotated as biological replicates in the first experiment (independent cultures from the same cell line) might be considered technical replicates in a design involving multiple individuals.

For example, if the goal is to generalize the results of the study across a population by sampling and treating cell lines from multiple individuals, each individual in the population sample would be considered a biological replicate. Meanwhile, replication of the treatment in each individual, such as separate treated cell line sub-cultures derived from the same person, would be considered technical replicates. This shift in annotation underscores the importance of accounting for the broader scope of variability when working with multiple levels of biological systems. Biological replicates represent independent units of biological variability, which could include different cultures, cell lines, tissues, or individuals, depending on the experimental context. In contrast, multiple samples derived from the same biological unit are conventionally considered technical replicates.

For instance, if the biological replicates are multiple liver samples from the same treated or untreated individual, the technical replicates would involve multiple rounds of sample processing, such as RNA extractions, from the same liver sample. On the other hand, if the biological replicates are liver samples from different individuals, then most would classify multiple tissue samples taken from the same liver as technical replicates.

The definitions of biological and technical replicates thus depend on the level of biological complexity being studied. However, in some experiments, multiple levels and types of replicates are involved, blurring the distinction between biological and technical replicates to the point where the specific annotation may be practically meaningless. Here's an example:

58

Consider a study involving treated and untreated primary peripheral blood mononuclear cells (PBMCs). PBMCs are a diverse group of immune cells, including lymphocytes, monocytes, and natural killer cells, that can be isolated from whole blood samples. These cells are commonly used in *in vitro* studies of immune function and response because they can be obtained relatively easily and are representative of systemic immune activity. In a PBMC treatment study, the aim might be to collect PBMC samples from different individual donors to investigate their regulatory and immune responses to specific bacterial infections. This could involve characterizing molecular differences in immune responses between individuals or examining how PBMCs respond to various types of bacteria. Replication could be considered at multiple levels, depending on the specific goals of the study, for example:

- **Repeated PBMC sampling from each individual**: To account for the variability in how PBMC samples from the same donor respond to the same treatment, we might collect blood samples on different days from the same individual. These replicate PBMC preparations would enable us to assess the consistency of the biological response within an individual.

- **Repeated treatment of each PBMC sample**: To account for variability in how the same sample of PBMCs responds to the same treatment, different aliquots of the isolated cells might be treated under identical conditions. These replicates would capture variability in the treatment response, independent of the variability introduced during PBMC isolation or sampling from the same individual on different days.

- **Sampling across multiple individuals**: To generalize findings, we would include samples from multiple individuals, adding another level of replication to the study. This level represents inter-individual variability in PBMC response, reflecting the natural differences in immune function and regulatory pathways between donors.

In this scenario, defining what qualifies as a "biological" versus a "technical" replicate becomes highly contextual. Moreover, in the various examples and scenarios of replication we mentioned above, we just focused on replicating the sample collection or intervention steps; we have not mentioned the purely technical replication related to sample processing for molecular assays, which can introduce additional nested levels of replication.

Indeed, the specific annotation of replicates is less important than the effective use of replication to achieve an effective design with respect to the goals of the study.

The case of tumor vs. healthy tissue

> **What I assume you already know**: In this chapter, I assume the reader has a basic understanding of concepts and workflows related to working with human tissue samples in functional genomics. While not strictly required to engage with this example, I also assume an implicit awareness of the ethical, regulatory, and practical considerations necessary for conducting experimental work involving human subjects and sampling human tissues. As in the previous case study, it is expected that you are familiar with RNA extraction from primary tissues and bulk RNA sequencing.

As we've learned, clearly defining the goal of a study is the first step in designing an effective experiment. When we wish to discuss gene expression studies comparing tumor and healthy tissues from humans, this can refer to a wide range of potential research questions, each with its own unique design requirements. For instance, consider these possible goals:

- **Within-individual comparison**: The goal might be to compare gene expression in tumor tissue versus healthy tissue from the same individual. This design focuses on identifying the specific molecular changes that occur as a tumor develops within one person. Because both tumor and healthy tissue samples come from the same individual, this minimizes the genetic and biological variability between samples. The primary advantage of this approach is that the differences observed are more likely to reflect true tumor-related changes rather than variability between different people. This makes it easier to detect subtle, disease-specific gene expression changes. However, the utility of the results may be limited to understanding how the tumor affects that particular individual, without necessarily generalizing across a broader population.

- **Between-individual comparison**: In this design, gene expression in tumors from one set of individuals is compared to healthy tissues from a different set of individuals. The goal here is to identify patterns of gene expression that are consistent across a population, not just within a single person. However, since the tumors and healthy tissues are sampled from different people, this approach must account for the added complexity of genetic, biological, and environmental variability between individuals. The challenge is to distinguish gene expression differences that are due to the tumor from those that arise naturally between people. This method is useful for generalizing findings to a broader population, but it requires larger sample sizes and more complex statistical controls to manage the variability between individuals. Thus, this design trades precision for generalizability, as it seeks to discover gene expression changes that are shared across different individuals.

- **Tumor type-specific comparison**: This approach might aim to compare gene expression between different types of tumors (e.g., comparing breast cancer tumors to colon cancer tumors) while including healthy tissue from the corresponding organ as a control. This study would seek to determine if there are tumor-type-specific gene expression changes or if certain gene expression patterns are shared across tumor types. This introduces complexity by requiring both healthy tissue and tumor tissue from different organs and tumor types. One would need to account for differences in cell composition across samples, among other factors.

- **Temporal comparison**: A fourth example might involve tracking changes in gene expression over time, from healthy tissue to tumor progression within the same individual. This approach requires multiple time points and introduces the challenge of understanding not only how gene expression differs between healthy and tumor tissues but also how it changes as the disease progresses.

Each of these study goals/designs presents unique considerations in terms of biological and technical replicates, the controls needed, and how to handle the variability between samples. The design choice depends on the goal of the study: understanding individual-specific molecular changes for personalized medicine, for example, versus identifying patterns that can be generalized across a population and may lead to the development of a treatment.

Gene expression levels in tumor vs. healthy tissue: study design

After considering the different configurations we mentioned above, another option emerges that combines the strengths of both within-individual comparisons and population-level analyses. In this formulation, we aim to examine both tumor and healthy tissue from the same individuals across a larger sample of individuals from a defined population. This approach offers the precision of within-individual comparisons while also providing broader insights from a population perspective.

Focusing on the practical aspects of the study, we will collect RNA sequencing data from small samples (~300 micrograms) from healthy lung tissues and corresponding lung tumors. Let us assume that informed consent was obtained and the study received approval from the Institutional Review Board (IRB). Before finalizing the study design, several key considerations must be addressed, particularly those related to tissue sampling.

It's important to note that this text will not explore the specifics of lung structure, function, or pathology; however, if you were designing this study yourself, a thorough understanding of lung biology and physiology would be essential. What we aim to emphasize here is how critical the tissue sample collection process is in a study of this kind.

The way samples are collected plays a major role in shaping the outcomes of the experiment. Poorly designed or inconsistent collection protocols can lead to confounders that persist throughout the study, rendering the results unreliable or even invalid. Careful attention must be given to the definition of "healthy" tissue, how far removed from the tumor it should be, and the composition of the collected samples. These considerations are central to ensuring that the study is both valid and replicable.

Let's explore this in more detail: When comparing tumor and healthy tissue within individuals, we need to carefully define what constitutes "healthy" lung tissue. The obvious choice for a sample from the tumor would be a section of the tumor mass itself. However, selecting the healthy tissue is more complicated. Ideally, the healthy tissue should be far enough removed from the tumor to avoid any influence from tumor growth, inflammation, or surrounding stroma. In that case, should we just sample the healthy tissue as far away from the tumor as possible?

Take a moment to think about this question and the properties of the samples that need to be considered.

One the one hand, we want to ensure that the healthy tissue sample is not affected by the tumor microenvironment. On the other hand, we don't want to sample the healthy tissue from an area so distant from the tumor that it may introduce unrelated variability in the lung tissue structure or cell composition. The lung consists of various cell types, including epithelial cells, immune cells, endothelial cells, and fibroblasts, each of which could have a different gene expression profile. Tumor tissue may have a vastly different cell composition compared to healthy lung tissue, which can strongly influence the observed gene expression data.

Given this complexity, a single bulk RNA sequencing sample from the tumor may obscure the distinct gene expression profiles of different cell types, such as malignant cells and the surrounding stroma. Similarly, healthy lung tissue may include a variety of cell types, such as alveolar epithelial cells and immune cells, which may also show distinct gene expression patterns. Undeniably, deciding how and where to sample primary tissues is a complex and nuanced challenge in experimental study design.

Let me take a moment to acknowledge that given these considerations, it is reasonable to ask whether bulk RNA sequencing is the most suitable technology for this study. **Single-cell RNA sequencing**, for example, offers the ability to characterize cell composition and cell states at high resolution in both tumors and healthy tissues, providing deeper insight into the various sources of variation in gene regulation. This approach may, in fact, represent the optimal choice for this study. However, for the didactic purpose of exploring sampling considerations and illustrating principles that can be generalized to other contexts, we will proceed with the discussion of this experiment under the assumption that **bulk gene expression data** are being collected.

Working with bulk data, given the complex structure of the tissues, raises an important question: is a single sample from the tumor and one from the "healthy" lung tissue truly sufficient? Tumors are often heterogeneous, containing a mix of malignant cells and stroma, including immune and connective tissue cells. Sampling one random section of the tumor mass might not fully capture the diverse gene expression landscape within the tumor environment. Similarly, healthy tissue could also benefit from being stratified into its key components, such as separating epithelial cells from immune cells, to better understand how each contributes to the overall gene expression profile.

Thus, we should consider whether to take separate samples from different compartments of the tumor, for example, the tumor mass itself and the surrounding stroma. The stroma plays a key role in tumor progression, influencing the immune response, angiogenesis, and other aspects of tumor biology. Sampling the tumor separately from the stroma could provide additional insights into how the tumor interacts with its environment and how various cell types contribute to gene expression changes.

Similarly, if feasible, separating the healthy tissue into its different histological components could provide a more detailed understanding of gene expression in both healthy and diseased conditions. As you can see, a study initially described casually as 'healthy vs. tumor tissue' can encompass many different possible scenarios, each with unique implications. Consequently, what might have seemed like a straightforward decision regarding tissue sampling has revealed itself to be a complex and critical choice that could profoundly affect the study's outcomes and effectiveness.

It is also important to realize that there is no definitive right answer when it comes to how we approach sampling in this example. The best choice depends entirely on the specific questions we want to ask. What's crucial is that we **make a deliberate and well-thought-out choice that aligns with the goals of the study**, while also remaining fully aware of the limitations inherent in each approach. Most importantly, we need to avoid introducing any confounders that could obscure the results.

For this study, we will opt for a bulk RNA sequencing approach where we sample, from each individual, both the tumor and healthy tissue several centimeters away from the tumor (to minimize the influence of the tumor microenvironment). We choose not to stratify the healthy tissue by cell type, focusing instead on a comparison between the overall tumor mass and a random nearby sample of putatively healthy lung tissue.

This choice allows us to answer a broad question: How does gene expression in the tumor compare to the general gene expression landscape of healthy lung tissue? While this approach doesn't allow us to explore the specific contributions of different cell types, it avoids the cost and complexity of single-cell analysis and provides a robust, population-level understanding of gene expression differences between tumor and healthy tissue.

A matched pair block study design

Let's reiterate: We will collect bulk RNA sequencing data from tissue samples obtained from individuals in a population sample, aiming to identify general patterns of gene expression differences between tumor and healthy lung tissue. When analyzing these data, we plan to focus on within-individual comparisons by examining gene expression differences between tumor and healthy lung tissue from the same person. By doing this across a population sample, we hope to identify consistent, generalizable patterns of regulatory differences between tumor and the corresponding healthy tissue. Tumor samples will be taken directly from the tumor mass, while healthy tissue will be collected from an area several centimeters away, far enough to avoid any influence from the tumor microenvironment. Each sample will be approximately 300 micrograms, enough to perform RNA extraction, library preparation, and sequencing.

This study follows a **block design**, where each individual serves as their own "block." Tumor and healthy tissues from the same individual are paired for comparison, and the data will be analyzed as paired samples (we do not discuss analysis in this primer, but you should look for paired-samples analysis in relevant textbooks). This design minimizes the impact of genetic and biological variability that naturally exists between individuals. By comparing tissues within the same person, we inherently control for confounding factors such as age, sex, environmental exposure, and genetics, focusing specifically on gene expression differences between tumor and healthy tissue. As a result, the block design improves statistical power by accounting for inter-individual variability, enabling the detection of subtle within-pair differences.

However, a key assumption in this design is that the "healthy" tissue is truly unaffected by the tumor. We are relying on the assumption that the tissue sampled several centimeters away from the tumor is genuinely healthy and not influenced by factors such as local inflammation, early tumor spread, or signaling from the tumor microenvironment. If this assumption is incorrect, our observations could be impacted by subtle tumor effects on the "healthy" tissue. In contrast, if we were to collect healthy tissues from entirely different individuals, we wouldn't have to worry about tumor influence on the healthy samples. However, that approach would introduce more genetic and biological variability between individuals, requiring a much larger sample size to account for these differences and achieve the same statistical power.

Thus, the *block design is the right choice* for this study because it allows us to control for individual variability while focusing on within-individual comparisons of gene expression. While we need to ensure that the healthy tissue is as unaffected by the tumor as possible, this approach offers greater statistical power and practicality than a design involving different individuals, which would require a much larger number of participants.

Let us assume that a power analysis (the details of which we will not cover in this primer) suggests that we need to collect paired samples from at least 20 individuals, each contributing both a tumor sample and a healthy tissue sample. These samples will be processed for RNA extraction, and we will collect bulk RNA sequencing data. As we move forward with the design, it's important to address several additional considerations that can significantly impact the results of our study. Avoiding confounders and minimizing variance are critical.

One key potential confounder is the order of tissue extraction. In clinical settings, tissues often remain at room temperature for a period after extraction, waiting to be either frozen by the clinical team or processed by the research team. If we have not explicitly accounted for this aspect of the study, it is possible that, without our realizing it, tumor tissue is consistently extracted before adjacent healthy tissue. As a result, the tumor sample may spend more time at room temperature than the healthy tissue sample.

Prolonged exposure to room temperature can lead to RNA degradation, which can alter gene expression levels. If degradation is consistently more severe in the healthy tissue samples than in the tumor samples, the observed differences in expression may reflect the extent of degradation rather than true biological variation. To mitigate this confounder, we can either ask the clinical team to randomize the order of tissue sampling or, more realistically given the practical constraints in the clinic, record the time each sample spends at room temperature before freezing or processing. Efforts should then be made to ensure that this time is similar across all samples. The effectiveness of these measures can be verified by directly testing for RNA degradation.

In addition to sample handling concerns during tissue extraction, we must also consider other biological and environmental factors, such as circadian rhythm. Gene expression fluctuates over the course of the day. While circadian rhythm is not likely to directly confound the comparison between tumor and healthy tissue within the same individual - because both tissue types are collected roughly at the same time - it could still introduce variability between individuals. If the time of day for sample collection varies across patients, it could obscure or exaggerate gene expression differences between tumor and healthy tissues when analyzing the population as a whole. More importantly, gene expression differences between healthy and tumor tissues might depend on circadian rhythm - an interaction effect - which could increase variability in the experiment and make it more challenging to detect true biological distinctions, even in a block design.

Unlike cell cycle variation, where allowing natural differences is generally acceptable, and sometimes the preferred approach, in this case, it is important to make an effort to collect all samples at the same time of day to minimize the impact of circadian rhythms and reduce unnecessary variability. This design choice, along with a detailed comparison between cell cycle and circadian rhythm as potential confounders, will be explored and discussed further in the third part of *Designing with Purpose*.

As with the first study we considered, using the cell line, we should be wary of introducing any number of potential confounding batch effects. For example, RNA extractions for tumor and healthy tissues should not be processed on different days or using reagents from different batches. Similarly, if sequencing is done in separate batches, tumor and healthy samples should be balanced with respect to the sequencing batch. Additionally, typical advice is to aim to replicate key experimental steps to help control for technical variability. We are going to discuss replication in more details the next section.

How can we improve the study if we had additional resources?

Considering the parameters of the study design we have discussed so far, let us explore how certain choices could further enhance the study's utility if additional resources become available. How should we allocate additional capacity? In other words, if we can process additional samples, and can access more patients, what strategy will be most effective in strengthening the study?

One option is to increase the number of individuals in the study. By adding more participants, we would improve both the power and the generalizability of our findings, enabling us to detect smaller gene expression differences that are consistent across a broader population. This would make the results more robust and applicable to a wider demographic. However, would this be the most effective use of additional resources?

Alternatively, we could focus on increasing replicates within the same individuals. Tumor tissue is inherently heterogeneous, meaning that different regions of the same tumor might exhibit distinct gene expression profiles. By collecting multiple samples from different parts of the tumor, as well as from the healthy tissue of each individual, we could better capture intra-individual variability. This would ensure that the differences we observe are not specific to a single location in the tissue, but instead representative of the tumor and tissue as a whole.

A third option is to replicate the RNA extraction and sequencing processes. By repeating these steps, we could control for technical variability, ensuring that any differences observed are truly biological rather than the result of errors introduced during RNA extraction or sequencing. While technical replicates might not directly address biological variability, they would improve the reliability of our measurements and reduce noise in the data.

Before reading further, take a moment to reflect: how would you approach the decision of selecting the best option? What considerations would guide your choice, and what information would you need to make it?

Now let's evaluate the trade-offs. The decision of where to allocate additional resources should primarily depend on where the most variability in the data is coming from. If individuals are highly variable in the way their tumors behave, adding more participants would be the most effective approach, as it would allow us to better capture the range of tumor behavior across the population. In this case, increasing the number of individuals helps us to more accurately estimate the variability between tumors and identify consistent patterns of gene expression.

On the other hand, it is possible that we are not effectively estimating the regulatory differences between healthy and tumor tissues because much of the variability lies within individuals, for example, due to differences in gene expression across various regions of the same tumor. In that case, collecting more samples from each individual would be more beneficial. This would enable us to capture the full spectrum of gene expression within a tumor and ensure that we are not missing critical information about tumor heterogeneity.

Finally, although this is unlikely in most experimental contexts involving tissues, if much of the variability arises from technical processes such as sample handling, freezing procedures, RNA extraction and quality, or sequencing, then replicating these technical steps would be the most effective strategy. This would help reduce noise introduced during the experimental steps and ensure that the observed differences between tumor and healthy tissue are truly biological.

Ultimately, the key is to allocate additional resources to replicate the biological or technical aspects that will most effectively help estimate and reduce variance. Whether that involves increasing the number of individuals, sampling more regions within each individual, or replicating technical processes depends on which source of variability is contributing the most noise to the data.

So, how do we know which option is most effective?

The answer can often be found by examining existing datasets or published studies. Consider the heuristic figure on this page, which shows boxplots of pairwise correlations (*y-axis*) of gene expression data across three classes of samples (*x-axis*): data from tissues sampled across different individuals, data collected from different tissue samples originating from the same individual, and data collected from separate RNA extractions of the same tissue sample.

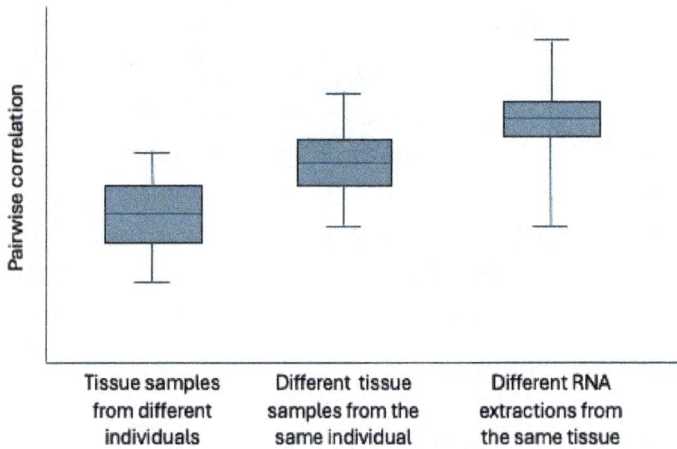

This figure illustrates a well-established pattern that applies to many, though not all, experimental contexts. Samples from different individuals generally display the greatest variation, reflected in the lowest pairwise correlations. Additionally, different tissue samples from the same individual exhibit more regulatory variation than independently extracted RNA samples from the same tissue. In this scenario, allocating more resources to include tissue samples from additional donors would likely be the preferred approach, as it addresses the factor contributing the highest regulatory variation.

Interestingly, it is worth noting that several studies have demonstrated that this cascade of regulatory variation does not always apply to cell lines. In the controlled *in vitro* environment of cell lines, combined with the short processing time of samples in the lab, RNA extraction itself can sometimes introduce more variability than sampling different aliquots of the same cell line.

When information about the relative contribution of variance from different sources is unavailable, conducting a pilot study becomes essential. Pilot studies will be discussed in the next chapter.

What have we learned from this example of study design?

The comparison of tumor and healthy tissue builds on principles we learned from the cell culture example, while adding complexities unique to primary tissues and across-individual comparisons. Like the first example, it underscores the importance of defining independent samples, mitigating confounders, and aligning design decisions with the goals of the study. Additionally, it highlights challenges unique to studies involving primary tissues and the complexities of population-level research, which must account for inter-individual variability.

This case reinforces several insights already gained from the previous example, including:

- **Defining the biological samples**: Just as we deliberated on culture size in the cell culture study, here we carefully defined "healthy" tissue to ensure it was unaffected by tumor microenvironment factors.

- **Avoiding confounders**: This case reiterates the necessity of maintaining consistency in sample handling and processing. In both studies, handling differences, whether in culturing time or tissue extraction, could introduce bias.

- **Balancing precision and generality**: As in the cell culture example, this study required balancing granular precision (e.g., sampling specific compartments of the tumor or healthy tissue) with the broader goal of deriving generalizable insights into the biological differences between tumor and healthy tissue.

Beyond these shared principles, the tumor and healthy tissue example highlights how effective study design evolves to address the increased complexity of population-level comparisons and heterogeneous biological systems. New considerations included:

- **Leveraging paired study design**: By pairing tumor and healthy tissue from the same individual, we minimized inter-individual genetic and biological variability. This block design allowed for more precise comparisons, isolating the effects of the tumor from unrelated differences across individuals.

- **Recognizing tissue heterogeneity**: The inherent complexity of primary tissues required us to be deliberate about sampling. Tumors are biologically heterogeneous, containing a mix of malignant cells and stroma. Similarly, healthy tissue consists of diverse cell types.

- **Minimizing environmental influences**: External factors, such as RNA degradation due to the amount of time a sample spent at room temperature can be a potential confounder. Ensuring consistent and timely sample processing was critical.

- **Generalizing across the population**: While the block design strengthened within-individual comparisons, extending findings to the population level required careful attention to sampling strategy (and statistical power). This trade-off between precision and generalizability is more pronounced in population studies than in experiments using cell lines.

- **Integrating biological and logistical considerations**: Practical constraints, such as attempting to ensure that tissues were collected at the same time of day to minimize circadian variability, underscored the interplay between biological rigor and experimental feasibility.

Designing with purpose (III): managing complexity in study design

In the simpler study involving a single cell line, we were able to provide clear answers to most questions about the study design, including what constitutes biological and technical replicates (though we did not explicitly use this terminology in the first example), how to avoid confounders, and other key considerations. However, as soon as we shifted to a slightly more complex study involving tissues sampled from multiple individuals, the answers became conditional, depending on specific circumstances. In some cases, practical considerations necessitated sensible compromises. This shift reflects a general truth about study design in genomics: as complexity increases, so does the need for careful evaluation of each decision. **Study design becomes less about following a strict set of rules and more about making thoughtful, informed choices that are aligned with the goals of the study.**

This doesn't mean that there are no wrong choices - there certainly are. A poorly designed study can lead to confounders that will compromise results, making it impossible to accurately answer the research question. However, in complex studies, there may not always be a single "perfect" solution. Instead, we must focus on designing the best possible study that can answer our key questions, while recognizing the limitations of what's feasible.

Sometimes, in the process of defining our research question, we may realize that the optimal study design to fully answer it simply cannot be achieved, given the constraints of available methods, samples, or resources. In these cases, it's not the study design that must be adjusted to fit the question, it's the question itself that may need to be revised. We must acknowledge the caveats inherent to the best study design we can realistically execute, and tailor our question accordingly. Often, the required compromises involve refining the scope of the question to limit the generality of the insight or accepting the possibility of alternative explanations that cannot be definitively ruled out. Following this approach allows us to make the most of what we can reliably measure, while being fully aware of the study's limitations.

At the heart of this process is the need to identify and avoid confounders. Whether it's variability in tissue collection, batch effects in sample processing, or differences in the way samples are handled across patients, confounders can easily undermine the validity of our study. The more complex the study, the greater the risk of introducing confounders that could obscure the real biological signals we're trying to detect. The goal of thoughtful study design is to strike the right balance between complexity and control, ensuring that we're answering the research question as accurately as possible without introducing bias that would distort the results.

Guidelines for effective replication

We should take a moment to consolidate what we've covered so far, particularly regarding how and when to replicate, before exploring additional study design examples. Consider the earlier examples: why did we decide not to synchronize the cell cycle, yet decided to collect tissue samples at the same time of day, in a sense, synchronizing the samples with respect to circadian rhythm? Why were we concerned that synchronizing the cell cycle might introduce another variable, while in the case of circadian rhythm, we chose to explicitly control for it?

More broadly, how can we determine whether a given parameter introduces a bias or systematic difference, rather than merely random variance? What are the general rules for choosing how to replicate? How do we decide when replication is necessary, and when it might not be required? Furthermore, in cases where replication is inevitable, how do we ensure that replication itself doesn't inadvertently introduce batch effects?

While not every scenario will have obvious and unambiguous answers, let's try to establish clear principles for effective replication in study design. Some examples in this discussion will repeat details we provided previously, but this time, armed with a slightly better overall intuitive understanding of study design properties, we will use these examples to derive general principles. Consider the examples of the cell cycle and circadian rhythm. Both are general properties that affect the regulation of thousands of genes in the cell, and both have the potential to interact with treatments or conditions of interest; this means that it is possible for the gene regulatory effect of a treatment (or condition) of interest to change depending on the cell cycle phase or time of day. However, key distinctions between cell cycle and circadian rhythm lie in their degree of specificity, timescales, and the practicality of controlling for their impact.

The cell cycle operates on a relatively short timescale, where even small differences in timing, such as 30 minutes to an hour, can lead to substantial changes in the expression of genes associated with the different cell cycles phases. The cell cycle is a property of individual cells, meaning that in an unsynchronized culture, we have a random collection of hundreds of thousands of cells, each at a slightly different phase of the cycle. The natural heterogeneity of cell cycle phases in an unsynchronized culture of hundreds of thousands of cells typically results in similar overall distributions of cycle phases across different unsynchronized cultures, making them comparable.

In contrast to unsynchronized cultures, imperfections in synchronization can introduce non-random variability over a short timescale. This occurs because the fraction of cells that escape synchronization no longer represents a random distribution of cell cycle phases; instead, they are either slower or faster to progress to the next phase. This can create a significant confounding factor where none existed prior to synchronization. To account for this variability, we will have to replicate the synchronization process itself, which truly defeats the point. For this reason, we chose not to attempt synchronization, which we would have had to replicate, and instead allowed the random natural variability in cell cycle phases to occur. We thus avoided introducing a systematic confounder while leveraging the inherent randomness of the cell cycle in large cell populations to achieve interpretable results.

Circadian rhythm, however, functions on a much longer timescale, with changes in gene expression accumulating over several hours. Attempting to randomize circadian rhythm effects in tissues is impractical in this context. Unlike the cell cycle, which is a property of individual cells within a culture, circadian rhythm is a property of an entire organism, a tissue sample, or a cell culture (circadian rhythm can also influences gene regulation in *in vitro* cell cultures, depending on the cell type). This means that there is no opportunity to assume a random distribution of circadian states within the sample that might average out across replicates. Instead, all cells in a tissue sample, or a cell culture, sampled at a particular time point will reflect the same circadian state.

Because sampling in the context of circadian rhythm typically involves a single parameter of interest – time - failing to control for this factor introduces substantial variability that cannot be mitigated by the inherent randomness seen with respect to cell cycle. To address this, we standardized the time of tissue sample collection by trying collect all samples at the same time of day. This approach effectively minimized the risk of confounding effects caused by circadian variability, resulting in more consistent observations. If we cannot synchronize the time of sample collection, we would have had to either balance it across conditions, which would result in the introduction of substantial variance to the experiment, or replicate it. (replication of this nature can easily be done in cell cultures but may be impractical if it involves repeated access to tissues from patients).

These examples illustrate a general principle: **the decision to replicate or control for a variable depends on whether the variability arises intrinsically within individual units (e.g., cells) or is uniformly applied to the entire sample.** Importantly, this distinction is based on the underlying mechanism driving the variability, not the variability itself. A variable that affects individual cells uniformly can still produce non-uniform variance.

When the factor driving variability is intrinsic and distributed randomly, as in the cell cycle phase, it can often be managed without explicit replication. Attempting to control for such variability (e.g., by synchronization) may introduce systematic biases that are harder to address. Additional examples for such factors include metabolic state, stress level, and in some contexts, cell composition. In contrast, when the factor driving variability impacts the entire sample, as with circadian rhythm, it should be intrinsically controlled or replicated. Other examples for such factors include case / control status, sample degradation, and practically all sample processing steps. These distinctions emphasize the importance of understanding the nature of the factor driving variability to determine whether standardization, explicit replication, or statistical adjustments are the most effective strategy for ensuring reliable and interpretable results.

The trick, of course, is determining both the amount and the nature of variability associated with each factor. Is the variability minimal or substantial? Is it random or consistent? Consistent variability can introduce bias into the experiment if not explicitly controlled. This is where the pilot study framework we mentioned multiple time becomes invaluable. By systematically comparing the variability introduced by different factors, pilot studies allow us to assess their impact and decide how to address them when designing the main study.

Interlude: The legacy of cDNA arrays and label bias

Before discussing pilot studies in the next section, I want to share a real example from the past where replication, through RNA label swapping in cDNA microarray experiments, unintentionally introduced additional variance that was not present originally.

Spotted cDNA arrays, which are no longer in use, were one of the earliest technologies for high-throughput gene expression analysis. These arrays consisted of thousands of DNA fragments ("spots") printed onto a solid surface, typically glass slides. RNA from biological samples was reverse transcribed into cDNA and labeled with fluorescent dyes before being hybridized to the spots on the array. The relative fluorescence intensities were then measured to determine gene expression levels.

The most commonly used labels were the **Cy3** and **Cy5** dyes, which emit green and red light, respectively. In a standard experimental setup, two RNA samples were compared directly through hybridization to a single cDNA array. One sample would be labeled with Cy3, and the other with Cy5, and both would be competitively hybridized to the same array. The fluorescence ratio of Cy3 to Cy5 at each spot provided a relative measure of the gene expression difference between the two samples. In a single comparison between two samples, a label swap replicate was necessary because the properties of Cy3 and Cy5 dyes, such as their binding efficiencies, fluorescence intensities, and rate of decay, were slightly different, introducing systematic bias that could only be balanced by performing the comparison with the labels reversed ('swapped') in a second replicate hybridization.

When a study involved a large number of samples that were hybridized to cDNA arrays, a typical experimental approach was the **reference study design**, where one sample (the "reference") was labeled and hybridized alongside each of the test samples. The reference sample was the same across all arrays, allowing for the comparison of test samples indirectly through their fluorescence ratios relative to the reference. This design helped standardize measurements across arrays and experiments (a 'reference-based incomplete block study design' for the statisticians among us).

However, an important factor, **label bias**, was often overlooked or misunderstood in early reference design studies. As we mentioned, Cy3 and Cy5 dyes differed in their fluorescence properties, leading to systematic differences in the measured signal that were independent of the true gene expression levels. Following the standard practice of label swap in two-sample experiments, some labs attempted to mitigate the label bias by using **reciprocal label swap** in the context of the reference design, essentially replicating the entire study with a flipped dye assignment.

This was entirely unnecessary. In a reference design setup, all test samples were labeled with the same dye, allowing them to be compared to each other on equal footing. In this setup, the investigator was not primarily interested in the direct comparison between the test samples and the reference, which was the only comparison affected by the bias associated with dye assignment.

If resources allowed for replication of the entire experiment, avoiding label swaps would have been the far better choice. Without label swaps, replication enhances the precision and power of the experiment by increasing the amount of data available to estimate gene expression differences, without introducing additional variables. In contrast, swapping labels in a replicated reference design introduced a new variable, the dye assignment, which created an interaction effect between the dye and the experimental conditions. This variable, at best, added unnecessary variance, and at worst, could confound results and reduce the overall interpretability of the data.

Designing a pilot study

When we don't have a clear understanding of which experimental step or process introduces the most variability, and the nature of that variability, it becomes challenging to design an effective study and decide where to allocate resources for replication. Without this information, we risk making decisions that either fail to address the primary sources of noise or unnecessarily replicate processes that don't contribute much variability. A well-executed pilot study can help clarify these uncertainties.

In many experiments, we also face practical limitations that require compromises in the study design. These compromises can sometimes introduce batch effects and confounders, but our goal is to select the confounder that is least likely to contribute meaningful variability. For example, consider a large-scale tissue culture experiment where space is limited, and we can't fit all cultures into a single incubator. In this situation, we need to decide whether it is better to carry out the entire experiment simultaneously and split the cultures between different incubators, or to perform the experiment at different times, using the same incubator each time. In either case, the experiment would be conducted in batches, but the key is to determine which option, different incubators or different days, introduces less variability.

Regardless of which option we choose, we would balance the experimental conditions (e.g., treatment vs. control) across batches to avoid confounding the batch effect with the parameter of interest. However, it is still essential to minimize batch effects by selecting the approach that contributes the least noise to the data. A pilot study can help us determine the optimal choice. In a pilot study, we would design a small-scale experiment to directly compare the variability introduced by using different incubators simultaneously versus performing the experiment over different days in the same incubator.

The key to this design is that all other variables must be held constant, so that the only source of variation we are assessing is the use of different incubators or culturing the samples on different days. To achieve this, we would set up replicate cultures that are identical in every respect - same type of cells, same media, same handling procedures, and ideally processed by the same individual. For the "different incubators" condition, we would divide the cultures evenly between two or more incubators, but all cultures would be treated simultaneously. For the "different days" condition, all cultures would be processed in the same incubator but staggered over different days.

The next step would be to collect data on the outcome of interest, such as gene expression levels or any other relevant biological readout, across all replicates in both conditions. The focus here is on estimating the variance associated with each condition. When analyzing the data, we would compare the variation observed across replicates between the different incubators and between the different days.

There are two general scenarios that can arise. The first is that factors such as different days or different incubators introduce mostly random variability, essentially a random increase in noise across the experiment. The second scenario is that these factors cause specific and consistent differences in gene expression, such as distinct gene expression patterns between cultures incubated in one incubator versus another. In principle, **we will almost always prefer conditions that introduce random variance (noise) over those that result in specific gene expression differences.**

Random variance can be managed by increasing the number of replicates. At worst, it adds noise to the experiment and reduces power. In contrast, conditions that cause specific gene expression differences can easily introduce confounding effects, even when the parameter of interest (e.g., treatment) is balanced across those conditions. This is because there is always a risk of a consistent interaction effect between the treatment and the condition (such as the incubator or day), making it difficult to attribute the observed effects solely to the treatment.

To evaluate the variability introduced by different experimental conditions, we can begin by analyzing and modeling the data. One common approach is to use an analysis of variance (ANOVA), which allows us to compare the variance between different groups, in this case, incubators or days, and determine whether there are significant differences in variability across these conditions. ANOVA works by partitioning the total variance observed in the data into components attributable to different sources, such as between-incubator variation and within-incubator variation. To assess whether the differences in variance are statistically significant, we can use an F-test, which compares the ratio of variance across groups. In our context, the F-test would allow us to determine if the variability introduced by using different incubators or different days is larger than what we expect when samples are process on the same day <u>and</u> in the same incubator.

While ANOVA and the F-test provide formal statistical analysis of the data, in this primer we try to avoid formal mathematics and statistics and provide more intuitive explanations.

To do so in this case, we can use a visualization technique that allows us to compare the variability across conditions and develop a sense of whether the differences are random or systematic. Let's assume we collected gene expression levels as part of the pilot; we can plot and compare the **ratios of gene-specific expression levels between different pairs of replicates cultured in different incubators.**

Specifically, as a first step, we are going to plot a comparison of the *ratios of gene expression* across replicate samples cultured in different incubators. Suppose that replicates 1 and 2 were cultured in one incubator and replicates 3 and 4 were cultured in a different incubator. On the Y-axis, we will plot the ratio of gene expression between replicates 1 and 3, and on the X-axis, we will plot the corresponding ratio of gene expression between replicates 2 and 4. Essentially, we are visualizing the correlation of effect sizes that result from culturing different, otherwise identical replicates, in different incubators.

Before looking at the figure on the next page, pause for a moment to think about what the data should look like under the different scenarios: one in which the incubator effect introduces mostly noise, and another where it results in specific and consistent changes in gene expression levels.

In the figure on this page, we plot the gene-specific ratio of expression levels between samples that were cultured in different incubators; on the *y-axis* the gene-specific expression ratios between samples 1 and 3, and on the *x-axis*, the ratios between samples 2 and 4. The axes have no values noted, but the standard approach is to plot the log-transformed ratios, such that zero (the origin) means that the expression levels across samples is identical (ratio of 1, or a log-transformed ratio of 0). The blue ovals represent the density distribution of gene-specific expression log-transformed ratio values for all genes (typically ~15,000).

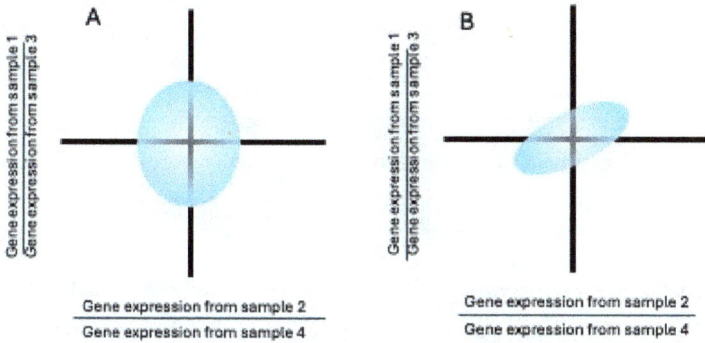

If culturing in different incubators introduces random variability in gene expression, essentially resulting in noise, we expect the effect sizes (i.e., the variability) to be mostly uncorrelated across repeated measurements. In this case, when comparing the differences in gene expression between pairs of samples cultured in different incubators (i.e., comparing the ratios of expression measurements), we would expect to see low correlation of effect sizes. This is illustrated in figure A, where there is variability between samples cultured in different incubators (the log-transformed expression ratios, or effect sizes, are not zero), but the gene-specific expression variability between replicates 1 and 3 is uncorrelated with the variability between replicates 2 and 4.

In contrast, figure B shows a clear correlation of effect sizes associated with culturing in different incubators, indicating a consistent effect. In the scenario plotted in figure B, the differences in gene expression between replicates cultured in different incubators are systematic and not random.

I should note that a complete lack of correlation, as depicted in Figure A for illustrative purposes, is expected only when the tested parameter does not specifically contribute to variance and the observed noise arises inherently from the experimental setup. In practice, whenever a tested parameter contributes to variance, some degree of correlation will always be observed in this plot. However, there will be a clear difference between parameters that primarily contribute noise (which will result in low correlation) and those driving specific and consistent changes in gene expression.

Regardless of whether culturing in different incubators results in random or consistent variability, we still need to ask: is the variability substantial or minimal? To answer this, we need a benchmark. Variance can only be assessed as high or low in relation to another factor, such as the effect of interest or another source of variability in the experiment. Without such a benchmark, it's difficult to judge the impact of any observed variability.

Recall that the purpose of the pilot study is to compare the variance associated with culturing across different incubators to the variance introduced by performing the experiment on different days. To do this, we need to conduct a simultaneous pilot experiment using "day" as the parameter of interest instead of "incubator." In this setup, replicates 1 and 2 are processed on the first day, while replicates 3 and 4 are processed on a different day.

This time, all replicates, regardless of the processing day, are cultured in the same incubator. As before, we will plot and compare the ratios of gene-specific expression measurements across days (i.e., replicates 1 over 3 and replicates 2 over 4). This time, however, we will present both pilot experiments (one in which variability is introduced by incubator and the other by day) on the same graph. The figure is on the next page.

Before proceeding, take a moment to review the previous figure (on page 89) and consider what it might look like if we overlay both studies on the same graph. Try to predict how the combined figure would appear if "day" introduces more random variance than "incubator," or if "day" results in random variance while "incubator" results in consistent effects.

Welcome back. In the figure on the next page, I show three scenarios. In all panels, blue ovals represent the density distribution of gene-specific log transformed expression ratio values between pairs of replicates cultured *in different incubators* (the same data shown in the previous figure), and red ovals represent the density distribution of gene-specific log transformed expression ratio values between pairs of replicates cultured *on different days*. The blue and red data points are independent data, collected in different experiments; we plot them on the same graph here for easier comparison. An appealing property of visualizing the data this way is that it enables us to draw intuitive conclusions, even without formal modeling or statistical testing.

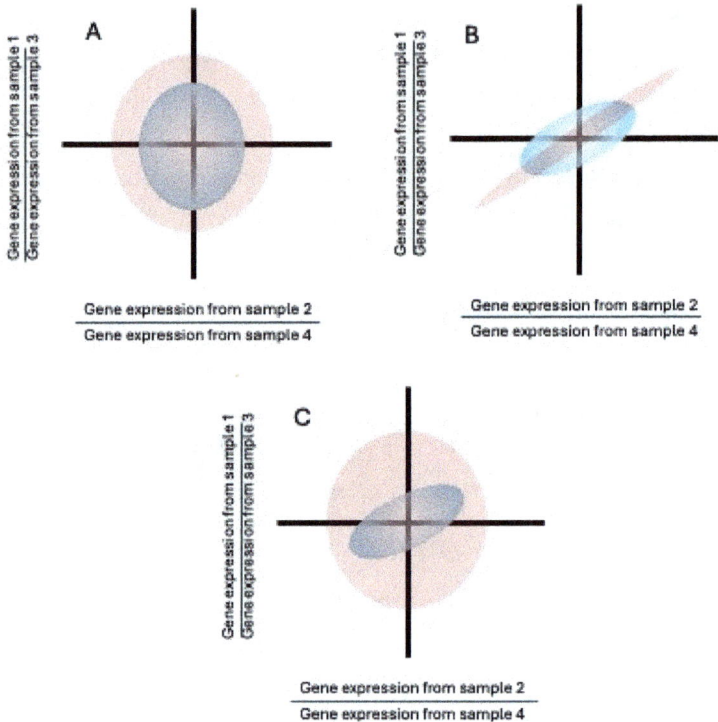

In panel A, we see that both "incubator" and "day" introduce random, uncorrelated variability. However, performing the experiment on different days (red oval) results in greater variability than culturing replicates in different incubators, suggesting that "day" contributes more random variance. In panel B, both "incubator" and "day" result in consistent, non-random differences in gene expression. However, the variation introduced by performing the experiment on different days is more pronounced, showing that the "day" effect has a stronger, systematic impact than the "incubator" effect. This scenario suggests that both conditions introduce non-random effects, but "day" has a larger impact on gene expression. **In panel C**, we find a mix of random and consistent effects. The overall variance due to "day" is greater, introducing a wider spread in the red oval, but this variance is uncorrelated; random. In contrast, the smaller variance introduced by "incubator" is consistent, as indicated by the correlation across the pairs of replicates cultured in different incubators.

This example highlights how a well-designed pilot study can guide decisions about where to allocate additional resources in the main experiment. By systematically comparing the variability introduced by different factors, such as incubator versus day, we can determine which source of variability has the greatest impact and how it should be addressed. For example, if the pilot revealed that day-to-day variability introduces more noise than incubator differences, the main experiment might prioritize performing all replicates on the same day while balancing other parameters, such as treatment, across incubators. Conversely, if only the incubator effect is systematic, the main experiment might prioritize using a single incubator, even if it requires performing the experiment over multiple days.

Importantly, this approach connects directly to the broader question raised earlier, in the context of the study of tumor and healthy tissues: how should additional resources be used to strengthen the study? Is the most effective approach to increase the number of individuals, collect more samples from within the same individual, or replicate technical processes? The insight gained from a pilot study helps identify where replication will be most effective in reducing variability. Without this guidance, we risk misallocating resources to replicate processes that contribute relatively little noise, while neglecting the factors that most significantly impact the reliability of our results. Ultimately, pilot studies provide a simple yet powerful framework for evaluating variability and benchmarking potential confounders, allowing us to design the main experiment in a way that maximizes statistical power and minimizes confounding effects. Armed with this insight, we can make informed decisions that ensure the study's results are both robust and generalizable.

The case of RNA decay experiment and gene expression after death

> **What I assume you already know:** In this chapter, I explain most relevant experimental concepts and make minimal assumptions about prior knowledge. However, as in previous chapters, I assume a basic understanding of the regulatory and practical aspects of working with human tissues, RNA extraction, and bulk RNA sequencing. Readers who are already familiar with RNA decay, RIN scores, and the extensive efforts empirical scientists undertake to prevent RNA degradation in their experiments may find it easier to engage with the material. Similarly, while the concept of time course experiments is explained in this chapter, having prior knowledge of it would make the material easier to follow.

Let's design a new experiment, one that will teach us an important lesson about an often-neglected empirical consideration of genomic study design. In fact, there's a good chance you've never explicitly encountered this critical aspect of functional genomics study design, even though it is inherently present in any genomic study you have conducted or read about. This time, our focus is on the process of natural RNA decay. We need to begin this discussion with a bit of background:

Understanding how RNA decays over time provides important insights into both the biology of gene regulation and the technical aspects of experimental reproducibility. RNA decay is a fundamental process that affects the stability and abundance of RNA molecules in cells. It occurs in two main forms: **regulated decay**, which is directed by cellular machinery to modulate gene expression, and **spontaneous decay**, which arises from the inherent instability of RNA molecules in various environments.

Both forms are crucial in determining the RNA landscape within cells, but here we focus specifically on the **spontaneous decay** of RNA, which occurs in any biological sample, especially samples left at room temperature for extended periods of time.

A common measure of RNA quality in experiments involving RNA sequencing and gene expression analysis is the RNA Integrity Number (RIN). RIN is a numerical score that reflects the level of degradation in an RNA sample, with values ranging from 1 (highly degraded) to 10 (intact, high-quality RNA). This score is determined by assessing the ratio of ribosomal RNA (rRNA) peaks in an electropherogram, specifically the 18S and 28S rRNA, and the overall RNA profile, which provides an indirect measure of RNA integrity. In many studies, particularly those involving primary tissue samples, RIN (or an equivalent score) is used as a proxy for the duration the tissue remained at room temperature before processing. This is because the time spent at room temperature correlates with the extent of RNA degradation that occurs between tissue collection and RNA extraction.

When RNA samples have different RIN scores, which means that their degree of degradation is different, it can introduce variability in gene expression measurements, because degraded RNA may be underrepresented in sequencing data. This raises the question: **how does RNA degradation affect different RNA types and genes?** If certain types of transcripts consistently degrade faster than others, the differences in degradation across samples could affect the apparent comparison of gene expression levels between samples. If certain RNAs degrade more rapidly, for example, they may be underrepresented in the sequencing data.

It is therefore necessary to study RNA decay in detail, not only to better understand the factors that influence RNA stability but also to guide the design of experiments and data normalization strategies in studies where some degree of RNA degradation is unavoidable. Knowing whether decay rates are consistent across experiments or vary randomly can provide valuable insights into how to handle RNA samples, interpret results, and minimize the impact of degradation on experimental outcomes.

In the study we are about to design, we therefore aim to compare the spontaneous decay rates of different RNA types across different genes. The central question is whether all RNAs degrade at the same rate under identical conditions or whether some RNAs are more prone to faster degradation. Furthermore, if degradation rates differ across transcripts, we need to know whether these rates are consistent across samples and experiments or vary randomly.

The study design framework for an RNA decay experiment

To study RNA decay, we need to design a **time-course experiment**, a study where measurements are taken at multiple time points to observe how a process unfolds over time. This approach will allow us to capture the dynamics of RNA degradation, while a single measurement would provide only a snapshot and miss the progression of decay across different RNA types. A time course experiment enables us to determine whether certain transcripts degrade faster than others and whether these patterns are consistent across biological replicates or influenced by technical factors. For an effective time-course experiment, several key considerations must be addressed:

- **Time Points:** Selecting appropriate time intervals is critical to capture meaningful differences in RNA decay. For example, if RNA degradation occurs rapidly, intervals of 15–30 minutes might be appropriate. However, if decay is slower, larger intervals may suffice. If there is uncertainty about the decay dynamics, we might first conduct a pilot study or consult previous studies to determine the optimal time points. A poorly chosen time course risks missing important changes or introducing unnecessary complexity.

- **Biological Replicates:** Biological replicates, which in this context are samples from different sources (e.g., different individuals or tissues), are needed for measuring consistency in RNA decay rates across experiments. These replicates allow us to evaluate whether decay rates are inherently predictable or vary between samples due to biological factors.

- **Technical Replicates:** Technical replicates, such as multiple aliquots from the same RNA sample, will help account for batch effects and ensure that observed differences are due to biological phenomena rather than experimental variability. They also provide a way to assess the reproducibility of RNA sequencing measurements.

- **Controlled Conditions:** To isolate RNA decay, all other variables must be minimized. For this experiment, RNA is extracted from primary tissue that is incubated at room temperature for different specified lengths of time (the time course), during which RNA degrades. All variables other than the time interval must be controlled, so that any observed differences in RNA abundance are attributable to decay rates, not environmental variability.

- **RNA extraction and sequencing:** At each time point, RNA is extracted from the incubated tissue, and an aliquot of the extracted RNA sample is immediately frozen to halt further decay. This step preserves the RNA state at each interval and allows for later processing in a consistent manner. At the end of the time course, when all RNA samples are available, we will establish libraries and collect RNA sequencing data.

Let's summarize: Primary tissues provide a biologically relevant context where the RNA content reflects the natural state of a living organism. Before RNA extraction, the tissue sample will be incubated at a controlled room temperature to simulate typical handling conditions. Extraction of RNA will be performed at regular time intervals (e.g., every 30-60 minutes) and immediately frozen to halt further decay. All aliquots will then undergo RNA sequencing together, to quantify the abundance of RNA molecules from different genes at each time point. Below is a graphical representation of the study design:

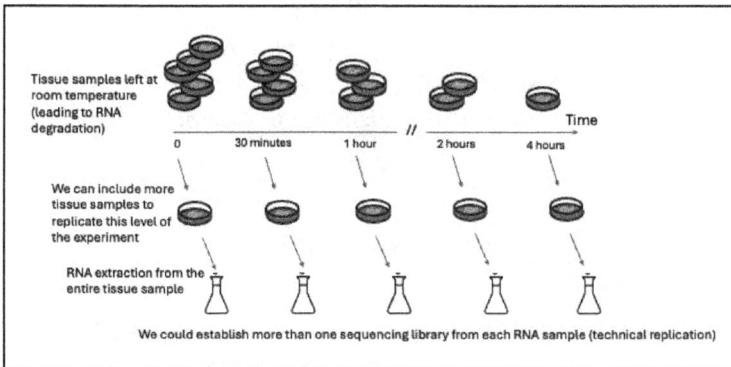

This experimental design allows us to examine the decay rates of RNAs across a wide range of genes. Biological replicates (not shown in the figure) help us evaluate whether decay rates are consistent across different tissues or individuals, while technical replicates (not shown either) ensure reproducibility and control for batch effects. If RNA decay rates are consistent for a given transcript across experiments, it suggests that the degradation process is predictable and can be accounted for in data analyses. On the other hand, if decay rates vary randomly, this variability could contribute to noise in gene expression studies, particularly when comparing samples with different handling times.

Identifying transcripts with inherently faster or slower degradation rates could also shed light on structural or sequence-based factors that influence RNA stability. By using RNA sequencing, we can achieve a high-resolution view of how different transcripts decay over time, capturing subtle differences in stability that may be missed with less sensitive techniques. This study not only addresses a basic question about RNA stability but also has practical implications for improving the reliability and reproducibility of RNA-based experiments.

Does the study design seem straightforward?

Let's assume we know the answers to all the usual questions about study power, such as how many samples to include, how many replicates are necessary at each level, how many reads to sequence from each sample, and so on. Let's also assume that we considered the usual confounders and either avoided them or balanced them with respect to the time-course. In other words, for the purpose of the following discussion, we are not concerned about cryptic batch effects or any other aspect related to confounders.

Given all the explicitly stated design parameters and considerations, there is one property of this study that I haven't mentioned yet, and it's absolutely critical! Take a moment to consider: how would you use the sequencing data to determine gene-specific RNA decay rates? If you can determine the answer to this question, it will reveal the critical component of the study design that has not yet been addressed.

Did you figure it out? The critical missing piece is **measuring or controlling for the RNA content before sequencing**. Without this step, the experiment lacks coherence; it simply wouldn't yield interpretable results. This requirement to control for RNA content introduces a complex and nuanced layer to effective study design. To illustrate this challenge, let's take a brief detour to examine an intriguing relevant real-world example: the claim that certain genes "wake up" after we die.

Do genes really "wake up" after death?

A few years ago, a study published as a preprint made headlines with the claim that certain genes "wake up" and become highly expressed up to 48 hours after death. The study used RNA sequencing data from mice to examine changes in gene expression over time following death. The authors collected tissue samples from various organs, such as the liver and brain, at multiple postmortem time points, including immediately after death and up to 48 hours later. Their analysis suggested a striking increase in the expression of certain genes, which they interpreted as evidence of postmortem activation. This conclusion captured public and scientific interest, as it implied that biological processes might persist or even emerge long after death.

However, the study's findings were based on a flawed understanding of the implications of standard RNA sequencing study design, and the biological processes occurring in postmortem tissues. After death, cellular processes cease, and RNA, an inherently unstable molecule, begins to degrade. The degradation process is not entirely random; some RNA fragments are more stable and persist longer, while others degrade more rapidly.

In their study, the researchers used a standard RNA sequencing workflow, which begins by extracting RNA from each tissue sample. Even though RNA is degrading over time, the researchers ensured that similar amounts of RNA were sequenced from each sample. This is a common practice in RNA sequencing studies, implemented to facilitate comparisons across samples. However, this practice assumes that the overall RNA quality is similar across samples. In the case of the time-course postmortem experiment, as RNA degradation progresses, the total pool of RNA becomes dominated by fragments from genes whose transcripts degrade more slowly.

This is quite a subtle study design issue.

As we discussed previously, RNA sequencing measures relative RNA abundance, not absolute gene expression levels, and as degradation progresses, certain RNA fragments may be disproportionately sequenced, creating an illusion of increased expression. To understand why this happens, imagine two transcripts: one that degrades rapidly and another that is much more stable. Immediately after death, both transcripts may be equally abundant. Over time, as degradation progresses, the unstable transcript will disappear more quickly, leaving the relatively stable transcript to dominate the RNA pool. When similar amounts of RNA are sequenced from each time point, the stable transcript will account for a larger proportion of the reads in the later samples. This proportional increase will be interpreted as higher relative expression level for the stable transcript, even though its absolute amount has decreased due to degradation.

The study design did not adequately account for the postmortem changes in overall RNA quality and quantity. As tissues were sampled at progressively longer intervals after death, the observed "increases" in gene expression were likely artifacts of slower RNA degradation rather than genuine biological activity. The appropriate controls, such as RNA integrity and quantity assessments, were not explicitly accounted for in the analysis. The failure to consider how degradation affects sequencing data, led to conclusions that were biologically implausible.

What the researchers observed as "gene activation" was actually a reflection of differences in RNA decay rates rather than genuine biological activity. Genes with slower-decaying transcripts appeared more highly expressed in later postmortem samples, simply because their RNA fragments were degrading slower than average. Without considering and modeling appropriate controls to account for overall degradation effects, this artifact can create the illusion of increased gene expression over time.

This study serves as a cautionary example of how experimental design and data analysis can fundamentally shape the conclusions. The idea of genes "waking up" after death is fascinating and thought-provoking, but the evidence presented in the preprint does not support this claim. Instead, it highlights the importance of thoughtful study design and careful data interpretation. By understanding how RNA behaves after death and incorporating appropriate controls, researchers can avoid being misled by artifacts and uncover the true biological processes at play. This story serves as a reminder that while advanced technologies like RNA sequencing are powerful, their outputs must be understood and carefully analyzed to avoid drawing incorrect conclusions. The study design problem in this case here was ignoring, or not realizing, that the study involved an implicit *empirical normalization step.*

Let's take a moment to talk about normalization

Normalization is a fundamental step in the analysis of functional genomics data. It involves linear and non-linear adjustments to the distribution of values in the data (such as the distribution of RNA sequencing reads per gene) to align the data across samples. While this primer focuses narrowly on study design rather than data analysis, normalization, though typically discussed in the context of analysis, has important implications for study design. To address these implications effectively, we need to first understand what normalization is.

As with all other parts of this primer, we will not explore or demonstrate the specific mathematical procedures underlying normalization. Instead, we will approach normalization intuitively, focusing first on its principles and implications for analysis and data interpretation, and later, for study design. This way, we aim to provide a clear understanding of the role normalization plays without requiring familiarity with the underlying calculations.

When you stop to think about it, normalization - the process of shifting the distribution of values to align them across samples - is a surprisingly counterintuitive concept. The primary goal of most functional genomics studies is to compare data across different samples, such as different individuals, tissues, treatments, or species. Yet, after carefully designing a study and meticulously collecting data from each sample to enable meaningful comparisons, the very first step we take, before evaluating differences between samples, is to adjust all the values through normalization. *Doesn't that seem odd?*

This process creates a conundrum: before normalization, it is nearly impossible to make meaningful comparisons of data across samples, yet the outcomes of those comparisons become heavily dependent on the normalization procedure. This reliance places substantial weight on the choice of normalization method, as it can profoundly shape the analysis and interpretation of results, potentially introducing biases or masking genuine biological signals.

Far from being a routine or trivial step, normalization plays a critical role in shaping the results and interpretations of functional genomics studies. Overlooking the importance of normalization risks introducing significant errors, as improper application can distort the data and result in misleading conclusions (we will give an example of this below). Ensuring that normalization is done correctly is not merely important, it is vital to maintaining the integrity of the study.

So... why do we normalize?

Normalization is used to adjust for technical variation across samples, ensuring meaningful biological comparisons can be made. Its purpose is to account for a wide range of potential biases that can arise during data collection and processing. For instance, correcting for sequencing depth is a common example of normalization and represents a relatively straightforward form of standardization. When we normalize (standardize in this case) for sequencing depth, we rescale data so that all samples are evaluated on an equal footing despite variations in total read counts. Another example is to normalize data in order to account for differences in total RNA yield between samples. In studies involving tissues or cell types with varying transcriptional activity, some samples may produce significantly more RNA than others. Without normalization, this difference in RNA quantity could skew the results. By ensuring that gene expression measurements are calculated relative to the total RNA pool rather than the absolute amount of RNA sequenced, normalization corrects for these differences and allows for meaningful comparisons across samples.

While correcting for sequencing depth or RNA yield are examples of relatively straightforward standardization, normalization extends beyond these cases. It includes more complex non-linear adjustments to account for broader technical and biological variability, ensuring that observed differences reflect true biological variation rather than artifacts introduced by technical inconsistencies in the experimental workflow. With rare exceptions, which we will not discuss here, **normalization involves shifting the distribution of expression values across samples without altering the rank order of values within each sample**. This preserves the relative relationships of values among genes while correcting for biases introduced during sample processing and data collection.

An example of normalization involving a simple linear shift is illustrated in the figure on the next page. The figure shows the distributions of expression levels for all genes in two samples. The expression level of a specific gene is highlighted (represented as a circle) in both samples. Before normalization (left panels), the gene expression distributions in the two samples differ significantly. At first glance, this might suggest that nearly all genes are expressed at higher levels in one sample (plotted in blue) compared to the other.

However, this difference is unlikely to reflect a true biological effect and is more likely a result of a technical bias.

To address this, we normalize the data by shifting the distributions so that their means are aligned. This adjustment corrects for the technical difference without altering the rank order of the relative gene expression levels within each sample. In the top panels, the figure demonstrates a gene that appears to be differentially expressed between the two samples before normalization. After normalization, however, we see that the rank order of expression for this gene is consistent across the samples, indicating that it is not differentially expressed. Conversely, the bottom panels show an example where the raw expression values for a gene appear similar between the samples before normalization, suggesting no difference in expression. However, after normalization, the adjusted values reveal a difference in rank order between the samples, indicating that the gene is truly differentially expressed.

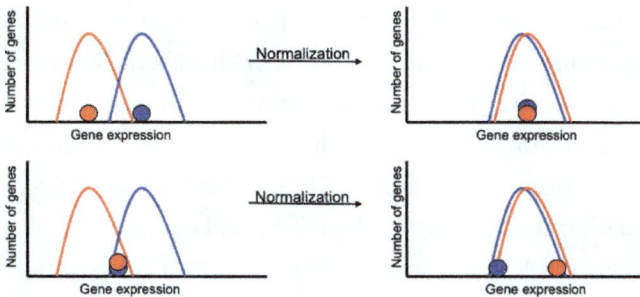

This example underscores how normalization corrects technical biases while preserving the relative relationships among genes, enabling accurate biological interpretation.

Normalization relies on two key assumptions, which I will explain in the context of RNA sequencing data. The first assumption is that the majority of genes are not differentially expressed between samples. This provides a stable reference for adjusting (normalizing) the data, as it presumes that most genes contribute similarly to the total RNA pool in each sample. Under this assumption, any overall differences in the distribution of gene expression levels across samples are attributed to technical variation, such as differences in RNA input or sequencing depth, rather than true biological changes. Normalization is designed to correct the overall distributions of expression values for these technical differences. However, if this assumption is violated, such as in experiments where most genes are genuinely differentially expressed, normalization can misrepresent the data, potentially under- or overestimating the expression of certain transcripts.

The second assumption is that variations in total RNA amount or sequencing depth are purely technical and do not reflect genuine biological differences. This assumption supports the idea that normalization removes unwanted variation, yielding a dataset that accurately represents biological changes. **However, in specific contexts such as postmortem studies or RNA decay time-course experiments, this assumption does not hold**. RNA degradation systematically alters the RNA pool over time, introducing biases that reflect the differential stability of RNA molecules rather than technical artifacts, and reducing the overall quantity of RNA in each sample. In such cases, overall observed differences in transcript abundance reflect biological effects that must be explicitly modeled and adjusted for, rather than eliminated through normalization. The biggest challenge with normalization procedures is that they can lead to subtle but significant errors in the interpretation of post-normalized data. Let me share with you a striking example of this, from the early 2000s, during the era when microarrays were widely used for gene expression studies.

Interlude: when normalization distorts the truth

We previously discussed microarrays in this primer, using the dye-swap as an example of a poor study design choice in a reference-incomplete block design. As mentioned, microarrays include gene-specific probes, and when labeled RNA hybridizes to them, the resulting probe-specific intensity is proportional to the expression level of the corresponding gene.

Around 2001–2002, commercial microarrays were designed primarily for a limited number of species, including human and mouse, with no arrays specifically designed based on the genomic sequence of any other mammal. Several research groups proposed the idea of using a 'human microarray' (that is, microarrays with probes specifically designed based on the sequence of human DNA) to perform comparative gene expression studies between humans and non-human primates, including chimpanzees, orangutans, and rhesus macaques.

In their published studies, these groups expressed an initial concern that sequence mismatches between the RNA from non-human primates and the human sequence-based probes on the array might attenuate hybridization efficiency with non-human RNA. If this attenuation occurred, the hybridization intensity of the human RNA would consistently appear stronger than that of non-human RNA, potentially leading to the erroneous conclusion that most non-human genes were expressed at lower levels than their corresponding human orthologs.

However, the authors went on to explain that among genes that were found to be differentially expressed between species, roughly half appeared to be highly expressed in humans, while the other half appeared to be highly expressed in the non-human primates. Based on this observation, the authors reassured readers that RNA from both humans and non-human primates hybridizes to the human arrays with comparable efficiency, showing no significant attenuation in intensity due to sequence mismatches.

The problem was that this observation was entirely an artifact of normalization! We can use the same figure we referenced earlier when we initially explained normalization (on page 107), to illustrate what happened in these comparative gene expression studies. In this case, let's assume the blue curve in the plot represents the hybridization intensity of human RNA, and the red curve represents the hybridization intensity of chimpanzee RNA. The reason the distributions are offset is that the hybridization of chimpanzee RNA is, in fact, attenuated due to sequence mismatches between the RNA from chimpanzee and the probes on the array, which were designed based on the human genome sequence. **This kinetic effect is simply unavoidable**.

When the investigators observed these differences, they followed the standard practice for handling overall differences in the distributions of functional genomics data: they normalized the data. Perhaps this was done out of habit, as normalization is commonly regarded as a standard step in functional genomics data analysis and is often applied without much thought. The result, however, as shown in the figure on the right panels, was that the distributions of hybridization intensities were artificially aligned and appeared symmetric, giving the impression that as many genes were highly expressed in humans as in chimpanzees. This erroneous symmetry obscured the true biological differences and introduced a false interpretation, driven entirely by the normalization process rather than the underlying data.

Normalization, in this case, masked the very real effect of sequence mismatches on hybridization kinetics and created a false impression that human and non-human RNA hybridize equally well to the human-designed arrays. The consequences of this error extended beyond a simple shift in the distribution of intensities, because it introduced many subtle non-linear inaccuracies that could only be resolved when multi-species arrays, that is arrays with probes designed based on the sequences of multiple species, became available.

We'll conclude the discussion of this example here, because the evolution of gene regulation in primates is not our primary focus. This example is meant to illustrate how analytical normalization, when applied without careful consideration of its underlying assumptions, can lead to incorrect inferences. Having a decent intuitive understanding of analytical normalization, we now finally, turn to the related topic of *empirical normalization* and its implications for experimental planning.

Empirical normalization is an important property of study design

Normalization is typically regarded as part of the analytical process, applied after sequencing to adjust the data for technical differences. It is distinct from empirical data collection, which focuses on generating raw, unbiased data from biological samples. As a result, normalization is not usually discussed in the context of study design. However, **normalization is also performed empirically** as a standard step in nearly all functional genomics workflows, even though it is rarely mentioned or explicitly discussed.

Empirical normalization, such as using similar amounts of RNA for sequencing, is a critical part of experimental workflows to ensure consistency across samples and to minimize technical variability, thereby laying the foundation for subsequent analytical normalization. It is important to appreciate that if genomic technologies could provide absolute measurements, we would not need to empirically normalize samples. However, because functional genomics, such as RNA sequencing studies, relies on comparisons of *relative measurements* between samples (as we previously discussed), empirical normalization is a fundamental step, without which meaningful comparisons would not be possible. Because the distributions of molecular data are often complex, without empirical normalization, differences in the initial amount of starting material could lead to non-linear distortions in functional genomics data, making it difficult to interpret observed differences between samples. By aligning the amount of starting material (equalizing RNA input, for example), variability introduced during the early stages of the experimental workflow is minimized, allowing subsequent analyses to better reflect genuine biological variation rather than technical artifacts.

I am sure you already see where we are going with this...

Despite its ubiquity, empirical normalization can also introduce challenges in certain contexts, such as decay or postmortem time-course studies. In these cases, RNA degradation alters the quality and composition of the transcriptome over time. As degradation progresses, the total amount of RNA per cell decreases, as molecules are broken down and lost.

To obtain similar amounts of RNA for sequencing at each time point in a decay or degradation time-course experiment, researchers must compensate for RNA loss by extracting the RNA from increasingly larger numbers of cells as degradation progresses. This approach ensures that the RNA input remains consistent across samples despite the ongoing breakdown of transcripts. However, this method of empirical normalization inherently alters the relative composition of the transcriptome. Transcripts that degrade at an average rate for the sample will appear to maintain stable relative quantities over time because the increased number of cells compensates for their overall loss. In turn, transcripts that degrade more rapidly than the average will become progressively underrepresented in the RNA pool, creating the false impression of decreased expression. At this point, this question should feel like a natural progression. However, let us pause briefly for emphasis: What happens to transcripts that degrade more slowly than average after empirical normalization?

The answer seems counter intuitive! Transcripts that degrade more slowly than the average, following an empirical normalization, will become proportionally more abundant in the sample, leading to the erroneous appearance of increased expression as time progresses.

Sounds familiar?

These are the genes whose expression seems to increase for up to 48 hours after death. This dynamic, driven by the need to standardize RNA input, demonstrates how elements of the empirical study design can interact with biological processes to shape the data. Empirical normalization, while essential for achieving comparable sequencing depths, is often implemented routinely, with little consideration of its underlying assumptions or potential impact. It is a step so ingrained in standard protocols that its implications are not always fully examined. In the context of decay or degradation time-course experiments, empirical normalization must be carefully incorporated into the study design to ensure that the results can be interpreted accurately.

How to control for the effects of empirical normalization?

In most contexts, the effects of empirical normalization are not a concern because we generally expect similar RNA yields across samples. A deviation from this expectation is assumed to stem from a technical issue and is initially addressed through empirical normalization, followed by analytical normalization of the data. However, in the context of degradation or decay time-course experiments, the situation is fundamentally different. By definition, we do not expect similar RNA yields from the same quantity of cells over time due to the ongoing degradation. Empirically normalizing RNA quantities in such cases introduces the artifacts described above, leading to misleading conclusions about gene expression changes over time.

One straightforward approach to account for the effects of empirical normalization in the context of degradation time-course experiments is to keep a record of the number of cells used for RNA extraction at each time point. By tracking this information, researchers can calculate the average RNA yield per cell. Over the course of the experiment, RNA yield per cell can be used to calculate the average rate of RNA decay and provide a baseline for estimating gene-specific decay rates relative to the mean decay rate.

This approach will allow for a more accurate interpretation of how individual transcripts are affected by degradation.

For example, let's assume we start with 1,000 cells at time zero and extract 1,000 nanograms (ng) of RNA, *giving an average RNA yield of 1 ng per cell*. At a later time point, due to degradation, we need 2,000 cells to extract the same 1,000 ng of RNA, indicating that the *average RNA yield has decreased to 0.5 ng per cell*. Now, consider two genes:

- *Gene A*: This gene decays at the same rate as the overall RNA pool, losing 50% of its abundance between time zero and the later time point. If the amount of RNA from *Gene A* in the entire RNA sample at time zero is 10 ng, its relative expression level would account for 1% of all transcripts in the sample extracted from 1,000 cells. As the sample degrades, the amount of RNA from *Gene A* decreases to 5 ng per 1,000 cells at the later time point, but because it degrades at the same rate as the overall sample, its relative expression is still 1%. When we process twice as many cells (2,000) to compensate for overall RNA decay in the later time point, *Gene A's* total contribution to the RNA pool remains proportional; its relative abundance stays constant at 1% in both time points. *Gene A's* apparent stability in the normalized data reflects its matching the average decay rate, not an absence of decay.

- *Gene B*: This gene decays more slowly, at half the average decay rate of the sample as a whole. Starting at 10 ng at time zero (1% of all transcripts in 1,000 cells), it decreases to 7.5 ng per 1,000 cells by the later time point. However, because **we processed twice as many cells** to compensate for overall RNA decay, *Gene B*'s total contribution to the RNA pool increases, and its relative abundance at the later time point rises to 1.5% (15 ng of all transcripts in 2,000 cells). This apparent increase in relative expression level is an artifact of empirical normalization. By keeping a record of the number of cells from which RNA was extracted at each time point, we can calculate *Gene B*'s true degradation rate, showing it decays 50% more slowly than the average transcript.

By using the known decay factor (based on average RNA yield per cell), we can determine the true dynamics of gene expression. *Gene A* is not stable, and *Gene B* is not increasing in expression; both transcripts are decaying, but at different rates. Recording the number of cells used to extract RNA at each time point allows us to calculate the specific decay rate of each gene, providing a clearer understanding of the biological changes occurring in the system.

Counting cells at each time point provides a theoretical solution for calculating gene-specific decay rates, but implementing this approach effectively poses significant challenges. Cell counts are prone to error, and the variability in each count can be substantial. To achieve reliable results, multiple counts per sample and multiple samples per time point are necessary to account for this variance, which can significantly increase the cost and complexity of the experiment. Additionally, the time required for cell counting introduces another layer of technical variability, as prolonged processing can lead to further RNA decay, compounding the issue. While the cell counting method is conceptually elegant, its practical execution can be difficult and resource intensive. An alternative approach that avoids these challenges is to use spike-ins, which offer a different strategy for addressing the limitations of empirical normalization.

Spike-ins are synthetic RNA or DNA molecules of known sequence and quantity that are added to biological samples during RNA sequencing experiments. These external controls, distinct from the sample being studied, are introduced at a stage where they remain unaffected by biological processes such as RNA degradation or changes in gene expression. By introducing spike-ins at a fixed amount across all samples, researchers can create a reference point to assess and normalize variation introduced during different phases of the experiment, including sample processing, RNA extraction, library construction, and sequencing.

Spike-ins are particularly valuable in experiments where RNA input or quality varies significantly between samples. Because their abundance is consistent and known, they provide a stable benchmark to evaluate how technical or experimental factors influence the observed transcriptome. By comparing the relative abundance of spike-ins across samples, it becomes possible to correct for systematic biases in the experimental workflows.

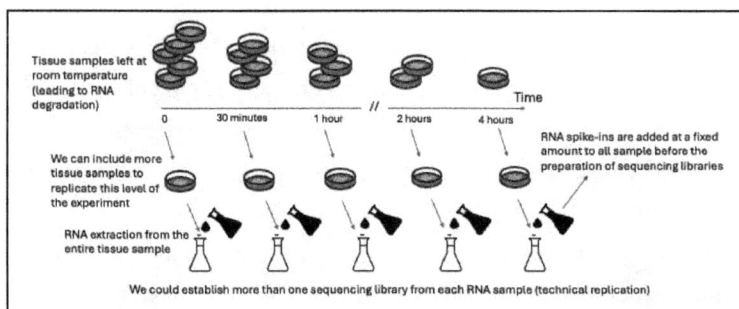

To implement this approach effectively in the context of an RNA decay time course, it is important to start with a comparable number of cells per sample at each time point. At first glance, this may seem like a challenge involving cell counting, as before.

However, it is significantly easier than matching the number of cells needed to achieve a consistent RNA yield. By starting with tissue samples of similar size at each time point in the decay experiment and adding spike-ins following RNA extraction (see figure), the entire sample from each time point can be processed consistently, such that RNA is always extracted from a similar number of cells. Replicates can help account for variation in the measured size of tissue samples. In most cases, this approach will yield more RNA than is needed for sequencing. Equal amounts of RNA can then be taken for sequencing, ensuring consistent input across all samples.

The relative proportion of spike-ins in each sample provides a direct measure of the degradation factor, allowing researchers to estimate the mean rate of RNA degradation without relying on precise cell counts.

For example, consider a scenario where spike-ins are added to each sample at a fixed amount of 10 nanograms (ng) per sample. At time zero, this represents 1% of the 1,000 ng of total RNA that was extracted from the entire sample (approximately 1,000 cells). At a later time point, RNA degradation reduces the RNA yield per cell, but the entire sample (again, approximately 1,000 cells) is still processed, yielding only 500 ng of total RNA. Since the spike-ins were added at the same fixed amount (10 ng), they now account for 2% of the total RNA pool. This proportional increase in the spike-ins indicates the mean degradation factor, serving as a benchmark to adjust gene-specific abundances and account for RNA decay.

By comparing the ratio of endogenous transcripts to the spike-ins, we can calculate the overall decay rate of the entire sample and, from that, determine gene-specific decay rates. This approach eliminates the need for precise cell counts for normalization, while still providing a robust method to account for variability introduced by degradation. Spike-ins, therefore, present a practical and effective alternative for addressing the challenges of empirical normalization in degradation time-course experiments.

What have we learned from this example of study design?

The RNA decay study builds on principles explored in previous examples while introducing new challenges and considerations unique to time-dependent experiments. We reinforced insights we gained from the previous examples:

- **Minimizing bias through careful design**: Rigorous attention to experimental setup and potential confounders ensures that observed RNA decay patterns reflect genuine biological processes rather than artifacts of the experimental design.
- **The value of pilot studies**: Conducting pilot experiments remains a key strategy to optimize parameters such as timepoints, replicates, and normalization approaches. This ensures that the experimental design is robust and effective.

New insights introduced in this example:

- **The analysis approach guides study design:** In this study, the chosen approach for calculating gene-specific decay rates played a pivotal role in shaping the experimental design. Defining the goal of estimating these rates clarified the need for precise control over the amount of starting material. This illustrates a broader principle: when employing dynamic treatments, the study design must be carefully aligned with both the research objectives and the analytical framework.

- **Empirical normalization as a critical property of study design**: This study underscores that normalization is not merely an analytical step but an integral part of experimental design.

- **The use of spike-ins**: The inclusion of spike-in controls provides a mechanism to correct for overall sample-level variability, making it possible to explicitly account for the effects of empirical normalization.

- **Timely recording of study parameters and metadata**: Accurate record-keeping, such as documenting the number of cells used for RNA extraction or the inclusion of spike-ins, is critical. Failure to record these details can render an experiment meaningless, as data lacking such parameters cannot be accurately interpreted or salvaged.

Designing with purpose (IV): batch effects and the importance of metadata

Let's pause before designing our next experiment to discuss batch effects directly. While this topic has been mentioned and addressed in examples throughout the primer, it merits explicit and detailed consideration. Batch effects, typically caused by non-biological variation introduced during experimental processes, can be subtle and unpredictable, yet their consequences can severely compromise the validity of functional genomics studies.

While it may not always be possible to eliminate batch effects entirely, thoughtful design can significantly reduce their impact. A substantial body of work addresses how to identify, correct, and account for batch effects during the analysis stage, particularly when batch effects are not fully confounded with the primary biological variables. However, this primer is not intended to address analytical approaches. Instead, we emphasize strategies for mitigating batch effects through careful experimental planning to reduce their occurrence from the outset.

A common notion about batch effects is that they can be ignored if they are minor. Setting aside the question of what is meant by '*minor*', this notion generally presumes that if the main effect of interest is strong, a minimal batch effect will not interfere with its detection. This reasoning is only valid under certain conditions. When a batch effect is completely confounded with the main effect, neither the strength of the main effect nor the subtlety of the batch effect matters. The confounding batch effect can render the entire study invalid.

When testing for a main effect, our goal is not just to detect it, but also to rule out alternative, trivial explanations that lack the depth or insight of the hypothesis we aim to support. Confounding batch effects can serve as precisely such trivial explanations. Even when batch effects appear minor, if they are fully confounding with the main effect of interest, and cannot be accounted for after data collection, they undermine the integrity of the study. In such cases, we cannot confidently attribute our observations to the intended main effect. Instead, the presence of an unaddressed, trivial explanation – the batch effect - diminishes the impact of the study, because it is fundamentally flawed. Without the ability to exclude confounding scenarios, the conclusions of the study lose their credibility and meaning.

Fortunately, many batch effects and confounders, including those highlighted in the earlier study design examples, can often be anticipated and either avoided or effectively managed. For instance, in studies with many samples, logistical constraints may prevent processing all samples in a single day, requiring them to be divided into batches processed on different days. This introduces a batch effect associated with the day of sample processing, but it is straightforward to address at the study design stage.

To prevent the processing day from confounding the primary parameter of interest (e.g., a treatment applied to half the samples), we can ensure that the processing day and treatment are balanced. This means that, on each day, we process an equal number of treated and untreated samples. This approach prevents the day or batch effect from becoming a confounder and ensures it does not bias the results. The processing day may still contribute substantial variance to the data (compared to a theoretical study that could have been completed in a single day). Depending on the power of the study, we may choose to use replication to explicitly estimate the variance associated with the processing day. However, importantly, by balancing the day and treatment, the processing day does not become a confounder of the treatment effect.

Many properties of study design can create a batch effect similar to 'processing day' and can be addressed in a similar way to avoid bias. Examples include differences in resources, such as the use of different kits or reagents, variation in equipment (such as sequencing machines, centrifuges, or PCR instruments), or environmental factors such as incubators with slightly different conditions (such as temperature, humidity, or CO_2 levels). In cell culture studies, variation in media composition or lot-to-lot variability in reagents can lead to confounding batch effects if different lots are used for treated and untreated samples. When different individuals collaborate on an experiment, the division of labor can introduce batch effects due sample handling.

In each of these cases, careful planning to balance the variable of interest ('treatment' in many of our examples) across potential batch-defining factors is essential to prevent confounding. Where perfect balancing is not possible or additional variability is anticipated, replication can be used to estimate and account for the variance introduced by these factors in subsequent analyses.

Some potential batch effects, however, are significantly more subtle to anticipate and plan for, making it important to maintain meticulous records of metadata. Comprehensive metadata allow researchers to identify and test for batch effects or major sources of variation, even with respect to parameters that were not explicitly balanced during the study design. Metadata should include detailed information about all aspects of sample collection, processing, and analysis. For example, the records should capture the date and time of sample processing, have a record of the individual performing the task, the lot numbers of reagents used, and the equipment or instruments involved. If multiple incubators or sequencing machines were used, the metadata should specify which samples were processed in each. Environmental conditions, such as room temperature at the time of sample processing, should also be documented if feasible.

Beyond technical parameters, biological metadata can be equally important. For instance, in a study involving patient-derived samples, metadata should include demographic and clinical details like age, sex, health status, and medication use, as these can introduce biological variability. In cell culture experiments, metadata might include passage number, cell density, and media changes. For time-course studies, precise timestamps for sample collection are essential.

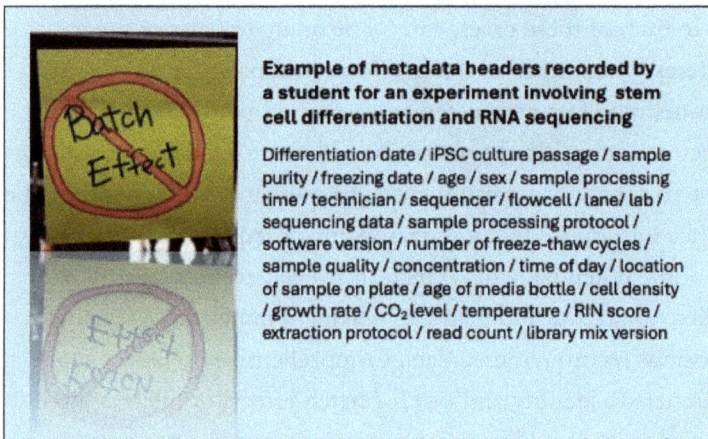

Example of metadata headers recorded by a student for an experiment involving stem cell differentiation and RNA sequencing

Differentiation date / iPSC culture passage / sample purity / freezing date / age / sex / sample processing time / technician / sequencer / flowcell / lane/ lab / sequencing data / sample processing protocol / software version / number of freeze-thaw cycles / sample quality / concentration / time of day / location of sample on plate / age of media bottle / cell density / growth rate / CO_2 level / temperature / RIN score / extraction protocol / read count / library mix version

A picture of a Post-it note I found one day in my lab (made by a former postdoc, Dr. Po-Yuan Tung).

Comprehensive metadata allow researchers to retrospectively assess whether unbalanced factors introduced systematic variation. For example, metadata might reveal a previously unnoticed issue whereby all treated samples were transported to the lab in a different type of container than untreated samples. Subtle differences in material composition or insulation between container types could affect temperature stability, ultimately altering sample quality. This would result in a batch effect fully confounded with the treatment variable, making it impossible to account for in the analysis (this is, unfortunately, a real example drawn from experience in my own lab). While such cases make the experiment challenging to interpret, identifying them is important for understanding the study's limitations and preventing false confidence in the results.

Most properties recorded in the metadata are not expected to contribute significant additional variance to the experiment; otherwise, they would have been explicitly addressed in the study design. Such properties are also unlikely to perfectly align with the parameter of interest (though this can happen unintentionally, as mentioned with respect to the example of sample container type). Consequently, variability in an experimental procedure discovered through metadata can often be statistically accounted for. For instance, when we review the metadata, we might discover minor differences in sample handling time, which may introduce variance but do not align perfectly with variables like treatment or condition. Having a record of the sample handling times allows for adjustments in the analysis to mitigate their impact, preserving the validity of the study. By documenting even seemingly minor details, researchers can identify unforeseen sources of variation and either control for them or flag them as potential limitations, ultimately improving the interpretability of the findings.

Unfortunately, when ignored, subtle unexpected batch effects can have significant consequences, sometimes undermining entire studies. Let's examine three real examples, each illustrating a different type of batch effect and its impact.

How batch effects misled the mouse ENCODE study

The Mouse ENCODE project set out to answer an important question: how does gene regulation differ between humans and mice? To address this, researchers measured gene expression in tissues from both species, using RNA sequencing to capture the activity of thousands of genes. The expectation was that, based on these gene expression data, samples would group together by tissue type. For example, human and mouse liver samples were expected to form a cluster based on gene expression data, reflecting their shared biological function, while human and mouse brain samples would form another cluster, and so on. This result would align with decades of evidence suggesting that tissues with similar roles exhibit comparable patterns of gene activity, even across species.

However, the Mouse ENCODE analysis produced a strikingly different result, illustrated in the figure on the next page. This figure, a heatmap, displays the correlations of gene expression data between all pairs of samples in the study. Human and mouse tissues are labeled with (h) and (m), respectively. The color gradient represents different levels of correlation, with red indicating the highest correlations. The samples are ordered and grouped based on their pairwise similarity matrix.

Notably, instead of clustering by tissue type as one might expect, the samples grouped by species. Gene expression data from human tissues grouped together, as did data from mouse tissues, regardless of tissue type. This unexpected result suggested that differences in gene expression were driven more by species-specific factors than by tissue function. If true, this would challenge a fundamental principle of biology: that homologous tissues in different species share conserved patterns of gene regulation.

A separate group of researchers reanalyzed the data to investigate the surprising result. This reanalysis uncovered a critical issue: a technical batch effect caused by the assignment of samples to flow cells on the sequencing instrument. The human samples had been processed on separate sequencing lanes from the mouse samples. Sequencing lanes, a component of the instruments used to read genetic material, can introduce systematic biases due to minor differences in their operation, such as variations in signal detection or reagent performance. Because the sequencing lane assignment

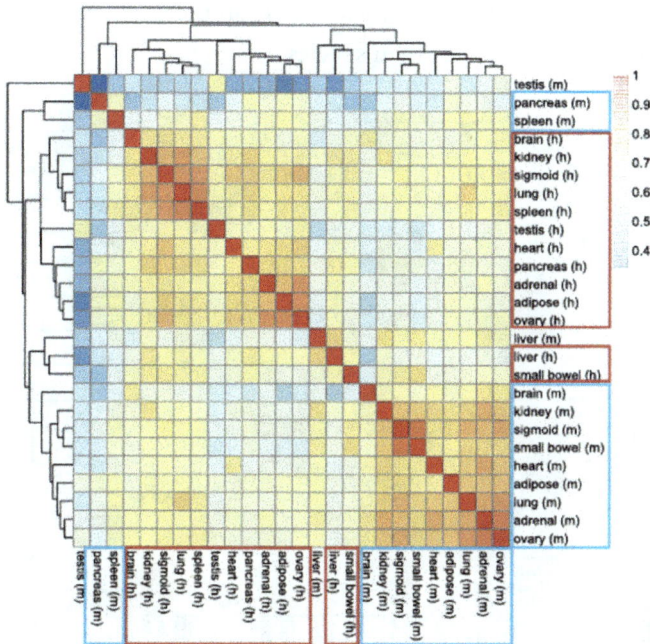

was completely confounded with species - human and mouse samples were processed on different lanes - it became impossible to distinguish whether the observed species-specific clustering reflected a true biological signal, or an artifact introduced by the sample assignment into sequencing lanes. The unexpected clustering result, combined with the complete confounding of species and lane, made it a likely possibility that lane-specific biases contributed to the observed patterns, artificially increasing the apparent differences between the species.

The complete confounding of species and sequencing lane posed a major challenge. Because sequencing lane and species were perfectly aligned, **any correction for lane-specific biases would also remove any real biological differences between species.** When statistical methods were applied to account for the batch effect, the clustering pattern changed dramatically. Nearly all of the samples now grouped by tissue, based on the corrected gene expression data, rather than species.

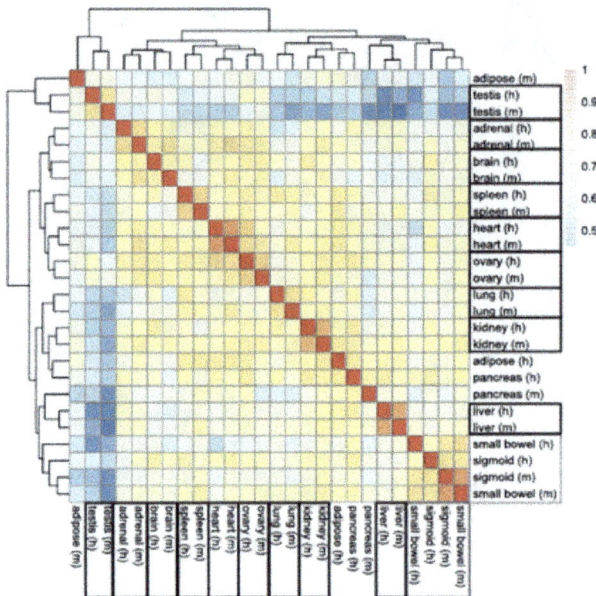

As can be seen in the figure on this page, human and mouse liver samples now formed a cluster, as did brain samples and other tissues. While this outcome appeared biologically reasonable, it also highlighted the problem with complete confounders: the correction removed both the technical artifact and any genuine species-specific signal, leaving only the tissue-specific signal intact.

This case illustrates the profound impact of complete confounders on scientific data. When a technical variable is perfectly aligned with the biological variable of interest, as it was here, no statistical correction can disentangle the artifact from the true biological signal. The intuitive clustering pattern seen in the data after correction can be misleading, because it does not necessarily indicate the recovery of accurate biological information. Instead, it may reflect the removal of both the confounder and the biological variable associated with it. The only way to adequately address this question now is to recollect the data using a proper and effective study design.

The illusion of precision of genomic sequencing

A team of researchers embarked on a large-scale effort to study human genetic diversity, using whole-genome sequencing to analyze DNA from populations across the globe. By combining data generated from multiple sequencing centers, the team aimed to map patterns in the genomic sequence that could explain phenotypic differences among populations and reveal new insights into human biology. The initial results appeared remarkable. Patterns in the data suggested meaningful genetic differences that seemed to align with population structures, hinting at exciting discoveries about the biological uniqueness of human groups.

However, closer scrutiny of the data revealed discrepancies that cast doubt on the original findings. Genetic variants that seemed common in one population, based on data from one sequencing center, were often completely absent in data from another sequencing center. At first, this was thought to reflect genuine biological variation, possibly due to regional genetic differences. Further investigation, though, showed that these patterns were artifacts of batch effects, technical inconsistencies introduced during the sequencing process in different centers.

The issue was further complicated by a geographical confounder. Many sequencing centers processed samples collected from nearby populations, leading to an overlap between technical biases and true regional genetic differences. This overlap made it exceptionally difficult to disentangle biological signals from artifacts caused by batch effects. What initially appeared as population-specific patterns of genetic variation was, in part, a reflection of differences in genomic sequencing between centers.

The sequencing process, contrary to what is often assumed, is not an absolute measurement of DNA but a probabilistic one. Sequencers don't directly "read" DNA bases (A, T, C, or G). Instead, they detect signals, such as tiny flashes of light or changes in electrical current, generated as the DNA is being 'sequenced'. The resulting signals are analyzed using algorithms that can infer the most likely sequence of bases, a process known as **base calling**. While highly accurate, base calling relies on probabilistic interpretations, meaning small differences in equipment calibration, chemical reagents, or the algorithms used can lead to some discrepancies in the results. Each sequencing center used slightly different machines, reagents, and algorithms to sequence the DNA. These differences, while subtle, influenced how the signals were interpreted, resulting in slight differences in base calling.

These probabilistic aspects of sequencing created the batch effects that misled the study. Coupled with the geographical confounding, the result was an inconsistent and biased picture of genetic differences across populations. The combination of technical and geographical factors made it extremely challenging to distinguish true biological variation from artifacts. This case highlights the unexpected vulnerability of genomic sequencing to batch effects. Although DNA itself is a fixed and unchanging code, the tools used to sequence it can introduce variability that obscures the truth. Addressing these challenges requires rigorous standardization of protocols across sequencing centers and careful quality control to identify and correct batch effects.

Another important safeguard is to balance the population origins of samples sequenced at each center, even if this requires shipping samples to more distant sequencing facilities. By ensuring that each center processes a mix of samples from diverse populations, researchers can reduce the overlap between technical and geographical biases, improving the reliability and accuracy of their findings. Without such measures, even the most advanced genomic studies risk being misled by artifacts of the technology rather than revealing the underlying biological reality.

Batch effects in a study of DNA methylation

Researchers studying the effects of environmental exposure on DNA methylation set out to determine how arsenic in drinking water might leave epigenetic marks on the genome. Using DNA methylation microarrays (of the few microarrays platforms still in use today), they analyzed samples from individuals exposed to high and low levels of arsenic. The initial findings were remarkable: more than a thousand DNA regions showed distinct methylation patterns between the exposure groups, strongly suggesting a link between arsenic exposure and changes in epigenetic gene regulation. These results pointed to plausible biological pathways that could underlie the increased risk of cancer and other diseases, following arsenic exposure.

However, deeper scrutiny revealed a critical flaw in the study design. The researchers had inadvertently introduced a bias by plating their samples in a specific order: samples from one exposure group were grouped together on the same batch of microarray chips, while samples from the other exposure group were plated on a different batch. This created a systematic confounding batch effect related to the order of processing of samples on the multi-chip microarray.

When the same DNA samples were reanalyzed with a corrected design, randomly distributed across the microarrays, the results told a very different story. Most of the previously identified methylation differences between the exposure groups disappeared. The conclusion was that in the original analysis, nearly all of the observed differences in methylation patterns were due to microarray-specific batch effects, not arsenic exposure. The confounding was so severe that even after using advanced statistical correction methods, the batch effects remained a dominant signal in the data, masking true biological differences. Only 25 methylation sites were found to overlap between the initial results and those obtained after randomization of the samples.

In the original study, the researchers compounded the batch effect issue by using pathway analysis software designed for gene expression data to interpret DNA methylation results. While this was an analysis choice rather than a study design flaw, it reflects a misunderstanding of the limitations and expectations of the chosen technology, a point highly relevant to our discussion in this primer ('what to expect from genomic technologies?').

Pathway analysis software is typically used to identify biological pathways or cellular functions that are associated with, and statistically overrepresented among, a list of annotated genes compared to the background of all studied genes. These annotated genes can, for example, represent those differentially expressed between conditions or, as in the study we are currently discussing, genes associated with differences in methylation between exposure groups. The specific software tool used by the researchers in this case was originally designed for gene expression datasets, where each gene is typically represented by a single probe or a small, consistent number of probes on the array. This uniform representation ensures that each gene contributes equally to the analysis, enabling meaningful comparisons across sets of genes.

DNA methylation arrays, however, are a fundamentally different technology. The number of probes per gene can vary widely, with some genes, particularly those frequently studied in diseases like cancer, being disproportionately represented. For example, genes such as *TP53* and *KRAS* may have dozens of methylation probes, while many other genes might have only one or none. This uneven representation results in greater statistical power to detect differences in methylation for genes associated with more probes compared to those with fewer probes. Recognizing and understanding this critical difference between gene expression and DNA methylation array platforms is essential for accurate analysis and interpretation.

When a pathway analysis approach designed for gene expression data was applied to identify enrichments of functional annotations in genes associated with differences in methylation, the underlying assumptions of the statistical analysis were violated. The uneven power to detect methylation differences across genes led to biased results, rendering the conclusions incorrect. This is a clear example of how a lack of clarity about what to expect from the technology, combined with insufficient familiarity with its properties, can result in misleading findings.

What initially appeared to be a groundbreaking study ultimately became a cautionary tale. The initial findings were invalidated by batch effects and the misuse of analytical tools that were incompatible with the chosen technology.

I selected these three specific examples because the flaws in their study designs, while avoidable, were subtle and unexpected. Indeed, we typically think of DNA sequencing as a categorical output rather than a probabilistic one, and it is not immediately obvious that systematic biases can arise from technical or procedural factors such as sample placement or processing order. These three examples demonstrate how even small oversights in study design can lead to devastating confounding effects. The key lesson is the necessity of proactive strategies to anticipate and mitigate batch effects. Achieving this requires deliberate incorporation of randomization (or balancing), replication, and careful metadata analysis at every stage, from experimental setup through to data interpretation. While these precautions may initially seem excessive for what appear to be minor details, they are essential for ensuring the reliability of results and the validity of insights.

The case of expression QTL studies

> **What I assume you already know:** In this chapter, we focus on study design principles for research involving genetic loci with regulatory roles. While an in-depth understanding of genetic associations and gene regulation is not strictly required, because the essential concepts are introduced in the chapter as needed, having some background knowledge can enhance your appreciation of the material. Specifically, familiarity with genome-wide association studies, promoters, enhancers, transcription factor binding sites, and the distinction between *cis-* and *trans-*regulatory elements, would provide helpful context.

This is the final detailed study design example in the primer. Each example we've discussed so far introduced unique challenges, added complexity, or highlighted different subtle aspects of study design that can have significant consequences if overlooked. In every case, the previous examples emphasized the critical importance of controlling for confounders and batch effects. The example in this section is particularly notable because of the way confounders interact with this type of study. In some ways, this design offers better safeguards against confounders and, perhaps more importantly, an intuitive method for identifying them. This is the framework of regulatory QTL studies.

A quantitative trait locus (QTL) is a specific locus in the genome where genetic variation is associated with inter-individual variation in a quantitative trait, such as height, blood pressure, or enzyme activity. These loci harbor genetic variants, such as single nucleotide polymorphisms (SNPs), that are inferred to influence measurable differences in a trait across individuals. When the trait of interest is gene expression, the loci associated with variation in the trait are referred to as expression quantitative trait loci (eQTLs). There are two types of eQTLs: **_trans_-acting eQTLs,** which influence both alleles of a gene and are typically located far from the regulated gene (sometimes on different chromosomes), and **_cis_-acting eQTLs,** which affect allele-specific gene expression and are generally located near the regulated gene. In this chapter, we focus on proximal eQTL associations, which are presumed to be _cis_ acting. The figure on this page illustrates an example of an eQTL. Each dot represents the expression level (y-axis) of the same gene measured in a different individual. These measurements are grouped by the genotype at a nearby locus (x-axis). The figure demonstrates an association between gene expression levels and the genotype at this locus.

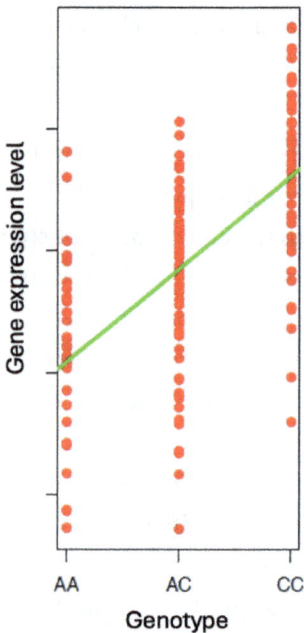

The eQTLs are inferred to impact the expression levels of specific genes, offering insights into the regulatory mechanisms controlling gene activity. The concept of mapping molecular _cis_ and _trans_ QTLs is not limited to gene expression, however. It extends to other phenotypes, including splicing (sQTLs), chromatin accessibility (caQTLs), and DNA methylation (meQTLs), to name a few examples. Mapping these regulatory QTLs provides a powerful framework for uncovering the genetic basis of molecular processes, advancing our understanding of how genetic variation drives cellular function, and ultimately, physiological phenotypes.

Regulatory QTL studies provide an intriguing example of how batch effects can introduce confounders. To illustrate this, we will focus on a specific example of an eQTL study. The goal of this study is to identify genetic variants that are associated with gene expression levels differences between individuals in a given species and tissue type. For simplicity, let us assume that the study involves human samples, targeting a specific tissue such as whole blood, and employs bulk RNA sequencing as the data type. Based on a power analysis, the researchers determined that a cohort of 100 individuals would provide sufficient power to detect thousands of *cis* eQTLs, namely, meaningful associations between genetic variants and inter-individual variation in gene expression levels.

The experimental design for such a study involves several key considerations and procedures to ensure robust and reproducible results. First, as this study aims to investigate associations between genotypes and gene expression, it is important to sample individuals from a single population. Population structure in the sample should be avoided, as it can give rise to spurious associations between genotype and phenotype in genetic association studies, including molecular QTL mapping studies. Beyond this brief note, the broader and critically important topic of population structure in genetic association studies (including regulatory QTL studies), and the corresponding analysis, are not discussed in detail in this book. Readers are encouraged to consult other texts for a comprehensive exploration of the subject.

Once the individual donors have been identified, researchers must carefully collect and store samples to preserve RNA integrity, adhering to standardized protocols to minimize variability introduced during handling. For example, blood samples should be collected in the same type of collection tubes, stored at consistent temperatures, and processed promptly after thawing to avoid RNA degradation. Collection timing is also important because gene expression levels can be influenced by circadian rhythms, as we discussed. To control for this, samples should ideally be collected at similar times of day across all participants. RNA is then extracted from the samples and prepared for sequencing, with quality checks at each step to ensure that degraded or low-quality RNA is excluded. Sequencing libraries should also be prepared using the same reagents and protocols, and individuals performing the tasks should be trained to follow consistent procedures. In a population study such as this, RNA sequencing is typically performed in batches, often dictated by the capacity of sequencing instruments.

To minimize batch effects confounding the results, samples should be randomized across sequencing batches. Since this is not a case-control study, there is no obvious way to 'balance' the study design, thus randomization is especially important to ensure that no single property, such as sex or age, is disproportionately represented in any one batch. Finally, let's assume that whole-genome genotype data are already available for all individual donors, enabling the analysis to identify associations between genetic variation and variability in gene expression.

Detailed metadata should be recorded for all aspects of the study, including sample collection times, processing steps, and sequencing batch information. This metadata facilitates the detection and correction of batch effects during data analysis, ensuring that unanticipated sources of variation can be addressed.

The orthogonal nature of eQTL associations

The eQTL study design appears relatively straightforward, adhering to the principles discussed thus far. However, one critical aspect warrants attention: the samples are randomized rather than balanced concerning unavoidable factors that could introduce batch effects. In prior experiments, a balanced design was often used to ensure that batch effects did not confound the primary effect of interest, such as a treatment or disease status. In the case of an eQTL mapping study, however, there is no specific treatment or property against which to balance the different batches. As a result, randomization becomes the only viable approach. What, then, are the potential implications of randomizing instead of balancing, and what strategies can be implemented to effectively control batch effects in this scenario?

For example, if we process and sequence 25 random samples per day, we create four distinct batches (the total sample size is 100). Randomization ensures that these batches are independent of systematic confounding by unpredictably distributing known properties, such as age and sex, across batches. **But what about the individual genotype?** It is unavoidable that any group of 25 unrelated individuals will share a unique subset of genotypes more frequently than other groups of 25 individuals, simply by chance. This occurs because any two unrelated individuals differ at millions of genotypes, but they also share a subset of genotypes by chance. When selecting 25 individuals at random, their shared genotypes can form a small but distinct combination of loci, just as a different group of 25 individuals would yield its own unique subset of shared loci. These overlaps of genotypes between individuals are not due to any systematic factors but arise naturally because of the immense number of possible genotypic differences among individuals. These unavoidable random patterns can lead to subtle genetic differences between groups - batches - of individuals.

When studying the relationship between gene expression and genotypes, batch effects can sometimes create misleading results. **This happens when changes in gene expression caused by the batch happen to line up with certain nearby genotypes**

that are shared by the individuals in that batch. Since any group of individuals is likely to share some genotypes by chance, the overlap of batch-specific genetic and gene expression patterns can create the false appearance of a connection between a genotype and gene expression where no biological association actually exists.

However, because genotypes are randomly spread across all the individuals in the study, only a very small number of these relationships will be affected by batch effects. Each proximal (putatively *cis-*) eQTL test considers one specific genotype and its link to the expression of one specific nearby gene, and these tests are mostly independent of each other. This means that while a few tests might show false results, the randomness of genotypes across the whole study ensures that most tests are not impacted by batch effects. (note: these considerations are somewhat different for *trans* eQTLs, which we are not discussing in this example).

Let us consider an example to illustrate why random genotype sharing and batch effects create false associations, but also why these false associations only affect a small number of tests. Let's use the same numbers as in our study: we are sampling 100 individuals, divided into four batches of 25 people each. Each batch is processed separately, and due to sample processing technical differences, gene expression measurements in the first batch (individuals #1–25) are slightly higher for a subset of genes. Batch effects, therefore, introduced technical variation in the measured expression level of some genes.

One of the important initial steps in data analysis is identifying and accounting for batch effects; as noted, this primer does not address data analysis. However, imagine that our batch correction is imperfect and leaves some residual impact; that is not uncommon. Now suppose that individuals in the first batch share 10 specific genotypes ***by chance***, which are not shared with the other 75 individuals in the study. If genes with higher expression in the first batch of samples, driven entirely by unaccounted residual technical factors, align with the shared genotypes (that is, the genotypes are located near the genes), false associations between the nearby genotype and gene expression will occur.

For instance, if *Gene A* is one of the genes with artificially higher expression in the first batch, and the same gene is also associated with one of the 10 genotypes shared by individuals in this batch, we will observe a false *cis* eQTL association for *Gene A* when we consider data from the entire study. This spurious association would incorrectly suggest that in our sample of 100 individuals, the genotype is driving the increased expression of *Gene A*, when in fact, the association is due to the technical batch effect and the random sharing of genotypes.

But here is where it gets interesting: these false associations will be quite rare! Consider *Gene B*, which also shows artificially higher expression in the first batch of sample due to the technical reasons we failed to account for. Unlike *Gene A*, *Gene B* is not associated with any of the 10 genotypes shared by individuals in the first batch (that is, these genotypes are not located near *Gene B*). Instead, the genotype associated with *Gene B* happens to be shared by a random collection of 25 individuals scattered across different batches in the study. Because these individuals were not processed together as a single group, the gene expression batch effect does not align with the individual genotypes, and no spurious *cis* eQTL association is observed for *Gene B*.

This example highlights a critical principle of effective study design: **the concept of orthogonal tests**. In the context of an eQTL mapping study, orthogonal tests refer to comparisons where the variables being tested, the genotypes in this case, are nearly independent from one another across the population. The vast majority of genotype combinations in the population are orthogonal to nearly every other combination, meaning there is little to no systematic overlap or collinearity among them. This inherent independence is what limits the impact of any technical batch effect to a small subset of tests in this type of study. Since batch effects are tied to specific groups or conditions, they can only align with a limited number of shared genotypes, leaving most tests unaffected. The lack of collinearity between most genotypes and batch effects ensures that spurious associations arising from technical artifacts remain rare.

In contrast to the eQTL framework, case-control or treatment-untreated studies often rely on a single primary contrast where all cases (or treated samples) are compared to all controls (or untreated samples). If a confounding batch effect coincides with this primary variable, such as if all treated samples were processed in one batch and all untreated samples in another, it can entirely dominate the analysis, as we have repeatedly discussed. This alignment introduces a systematic bias that affects the central comparison, making it impossible to disentangle true biological differences from batch effects. Unlike orthogonal tests, where confounding is distributed across many independent comparisons, the single contrast in case-control designs makes the study highly susceptible to confounders.

This is not to suggest that efforts to minimize batch effects in molecular QTL studies are unnecessary. However, the risk of confounders is generally limited, with batch effects primarily affecting overall variance and reducing statistical power.

Dynamic eQTLs present a different challenge

Dynamic eQTLs extend the concept of eQTL mapping by focusing on how genetic effects on gene expression vary across different biological contexts or conditions. Unlike static eQTLs, which are typically identified under uniform conditions and represent consistent genetic effects, dynamic eQTLs capture the ways in which these effects change across various biological and environmental dimensions. Dynamic eQTLs reveal not only the presence of genetic regulation but also its flexibility and context dependence, offering deeper insights into the complexity of gene regulation. There are different types of dynamic eQTLs, including:

- **Spatial dynamics (across cells and tissues):** Genetic effects on gene expression may vary between cell types or tissues. For example, an eQTL that influences gene expression in liver tissue might be inactive in brain tissue. Mapping dynamic eQTLs across tissues or cell types can help uncover tissue-specific genetic regulation, which is particularly relevant for understanding complex traits linked to specific organs or systems.

- **Temporal dynamics (across timepoints or developmental stages):** Genetic effects can also shift over time, particularly during developmental processes or aging. For instance, an eQTL active during embryonic development might no longer influence gene expression in adult tissues. Temporal dynamics in eQTLs provide critical insights into how gene regulation adapts during an organism's life cycle.

- **Response dynamics (to environmental exposure or treatment conditions):** Certain genetic effects only emerge or change in response to specific stimuli, such as drug treatments, stress, or infection. These response eQTLs are key to understanding how genetics interact with the environment to shape phenotypes and may uncover mechanisms of drug response or disease susceptibility.

Dynamic eQTL mapping marks a transition from perceiving genetic regulation as fixed to recognizing its context-dependent variability. By examining these dynamics, researchers can reveal genetic effects that are overlooked in standard static eQTL studies, providing a deeper and more nuanced understanding of how gene regulation influences complex phenotypes. The study design for dynamic eQTL mapping is more intricate than that for standard eQTL mapping due to the need to introduce a specific perturbation. However, the core considerations remain consistent across different types of dynamics. To illustrate these principles, let us focus on designing a study that examines temporal dynamics across a developmental trajectory.

Mapping dynamic eQTLs across a developmental trajectory

The experimental design for mapping temporal dynamic eQTLs requires additional considerations to effectively capture changes in genetic regulation throughout a developmental process. For this example, we will study the differentiation of stem cells into cardiomyocytes (heart muscle cells) using single-cell RNA sequencing data collected at multiple time points along the differentiation trajectory. We will use stem cell cultures that were previously established from 100 individuals from whom we already have whole-genome sequence and genotype data.

- **Sample preparation and differentiation protocol:** To ensure consistent results, the stem cell differentiation protocol must be meticulously standardized. Stem cells should be cultured under identical conditions, using the same growth medium, supplements, and differentiation-inducing agents. Care must be taken to minimize variability in environmental factors such as temperature, CO_2 levels, and humidity, which can influence differentiation efficiency and gene regulation. Samples should be collected from the differentiating cultures at pre-defined time points during the differentiation process, such as Day 0 (pluripotent stem cells), Day 3 (early differentiation), Day 7 (intermediate stage), and Day 14 (mature cardiomyocytes). These time points were chosen to represent distinct phases of the developmental trajectory.

- **Single-cell RNA sequencing workflow:** The same single-cell dissociation protocol should be used to isolate individual cells from all samples, with care taken to minimize stress-induced gene expression changes during the dissociation process. Once cells are collected, they should be processed promptly to preserve RNA integrity. Libraries for single cell RNA sequencing should be prepared using the same reagents and kits across all samples, and individuals processing the samples must adhere to standardized procedures. Single-cell RNA sequencing allows for capturing the heterogeneity of cell populations at each time point, making it ideal for studying developmental trajectories. To minimize batch effects, all samples should be processed together for sequencing to ensure consistency.

As always, meticulous metadata will be recorded for all samples and processing steps. Notably, we did not use technical replicates for the differentiation of the same individual, opting instead to allocate resources toward adding more individuals. This approach aligns with the principle of prioritizing additional samples where the greatest variance arises, as we discussed previously. Additionally, we opted to process RNA samples in batches immediately after the cultures were collected to prevent degradation. An alternative reasonable approach would have been to freeze the cultures at each collection time point and extract RNA in a single batch after all cultures were collected. Regardless of the RNA extraction design, all samples will be sequenced together to minimize batch effects associated with sequencing.

Once the data are collected, the analysis will aim to identify dynamic eQTLs, which are eQTLs where the association between a genetic variant and gene expression varies along the differentiation trajectory. For example, a genetic variant may show a strong association with the expression of a gene during early differentiation, in the cell types that are common on Day 3, but this association may diminish or disappear by Day 14. This dynamic variation provides insights into context-specific genetic regulation of gene expression.

Cryptic batch effects can undermine dynamic eQTL studies

By now, you're likely weary of hearing about batch effects, but this final example of study design was chosen precisely because it is susceptible to an unusual and subtle type of batch effect. So, one last time, let's explore batch effects in this experiment.

As we discussed, in standard static eQTL studies, the orthogonal distribution of genotypes in a population sample typically limits the impact of technical confounders, making false positive eQTL associations relatively rare. This principle holds true for eQTLs identified at any single time point or single context. However, does the same hold true for dynamic eQTL associations? What differentiates identifying a steady-state (static) association at a single time point from detecting a dynamic pattern? How might batch effects uniquely impact the identification of a dynamic process, where eQTL effect sizes vary across time points? Take a moment to reflect on these questions. Consider this as a clue: the type of batch effect we are about to describe has a unique characteristic: It is unavoidable and cannot be fully mitigated through study design.

Welcome back. This is a subtle and complex topic, and likely the most challenging to explain intuitively among all the subjects covered in this primer, especially without relying on formal mathematical treatment. Dynamic eQTL studies, such as those mapping changes in eQTL effect sizes along a developmental trajectory, differ fundamentally from standard eQTL studies in terms of susceptibility to batch effects. In our example, a dynamic eQTL refers to an eQTL with varying effect sizes across developmental time points. We mentioned the example of a genetic variant strongly associated with variation in gene expression at an early stage of differentiation but not at later stages.

In a standard steady-state eQTL study, batch effects can lead to false associations, but only under specific conditions. For a spurious standard eQTL to emerge, a batch effect must alter the expression of certain genes, and the genotypes near these genes must be, by chance, unique to the individuals in the affected batch. Such scenarios are rare, because efforts to control batch effects, both during study design and analysis, generally minimize their impact, and because - as we discussed - the random distribution of genotypes across the population reduces the likelihood of systematic overlap between batch effects on gene expression and nearby genetic variation.

Dynamic eQTL studies face a unique and more significant risk, one that cannot be fully avoided through study design alone. In the context of a dynamic study, batch effects can emerge from variations in how samples from specific donors progress through the developmental trajectory, respond to a treatment, or react to a perturbation. **In other words, batch effects in the context of dynamic eQTL studies can arise from genuine biological differences.** This is particularly confusing because, while the batch effect may reflect actual biology, it can still lead to the spurious identification of dynamic eQTLs.

In the example of stem cell differentiation, a batch effect could occur if cells from some donors follow a slightly different developmental trajectory than the cells from other donors. These biological variations in the differentiation process can lead to widespread changes in the expression of many genes at the timepoints where these differences occur. When genes influenced by these differentiation-related changes are also associated with genetic variants that, by chance, are specific to the group of donors whose cells underwent a different differentiation process (the 'batch'), a significant problem arises. In such cases, these genes can appear to have time-dependent *cis* eQTL effects, creating the illusion of dynamic eQTLs. While dynamic *cis* eQTLs associations are still orthogonal to each other, this type of batch effect can be quite widespread, as changes in the differentiation process (much like response to many treatments or perturbations) can impact the expression levels of hundreds or even thousands of genes simultaneously, some of which will be associated with nearby genotypes shared by the individual in the batch.

The batch effect in dynamic eQTL studies differs fundamentally from that in standard eQTL mapping because it cannot be avoided. In standard eQTL studies, batch effects typically arise from the technical grouping of samples, which can create differences in how subsets of samples are processed or handled. These effects can often be minimized through careful study design. In dynamic eQTL studies, however, the batch effect arises because individual donor samples respond differently to the perturbation itself. In our example, this manifests as some samples following distinct differentiation trajectories. This variability reflects a genuine biological process that differs among individuals, rooted in their inherent biological properties. Yet, these differences can result in hundreds or even thousands of spuriously identified dynamic eQTLs. No study design consideration can fully address this issue, even if we recognize it in advance, as it originates from the inherent biological characteristics of the donors.

Another key distinction between standard and dynamic eQTL studies is that identifying batch effects is considerably more challenging in dynamic studies. In standard eQTL studies, cryptic batch effects are often uncovered by examining metadata to find variables associated with significant variance or through analytical methods that estimate the influence of unmeasured factors. In dynamic eQTL studies, however, it is far more difficult to detect when certain individuals respond differently to the perturbation at specific time points. The biological 'batch' can often remain undetected until dynamic eQTLs are mapped, at which point the artifacts have already been incorporated into the analysis.

To illustrate how these effects can be identified, we must briefly explore an analysis approach. While this extends beyond study design, it is essential for a complete explanation and practical understanding of these challenges.

To detect spurious dynamic eQTLs caused by batch effects, we can leverage the principle that genotypes are distributed orthogonally across individuals. This means that for any given genotype, the group of individuals who share it will vary, and the overall pattern of genotypes associated with dynamic eQTLs should be unique for each individual. To explain this more clearly, consider the vector of genotypes associated with all identified dynamic eQTLs. For each individual, this vector represents the alleles (major or minor) they carry for these eQTLs.

For example, suppose 500 dynamic eQTLs are identified, each with a major and minor allele. Individual 1 might carry the major allele for the first 100 eQTLs and the minor allele for the remaining 400. Individual 2, on the other hand, could carry the major allele for all but eQTLs 100 through 180. Similarly, individual 3 might have a completely different pattern, such as the minor allele for the last 150 eQTLs and the major allele for the rest. These genotype vectors will vary independently across unrelated individuals, reflecting the orthogonal nature of unlinked genotypes.

If we compute pairwise correlations of these genotype vectors across the entire sample, we expect no systematic pattern; any two individuals should appear equally similar or dissimilar to each other. Visualizing these pairwise comparisons in a heatmap should show no discernible structure. This is illustrated in the **left panel** of the figure on the next page, which shows the pairwise correlation matrix of genotype vectors for *steady-state (standard) eQTLs* identified in a time course study with 19 individuals (the identifiers of the individuals are listed in the same order on the X and Y axes). The colors represent correlation coefficients, with darker shades indicating higher correlations. The figure on the left reveals no apparent structure, consistent with the expectation of random genotype distribution for the vector of standard eQTLs. This pattern confirms that these eQTLs are unlikely to have arisen from a biological batch effect.

The **right panel** of the figure shows the pairwise correlation matrix of genotype vectors for the *dynamic eQTLs* discovered in the same time course study. This time, one can detect a clear structure in the heatmap (highlighted by the blue squares). This structure indicates that the individuals within the clusters (squares) share many more dynamic eQTL genotypes with each other than with individuals outside the cluster. Such a pattern is expected when a biological batch effect impacts the perturbation or time course.

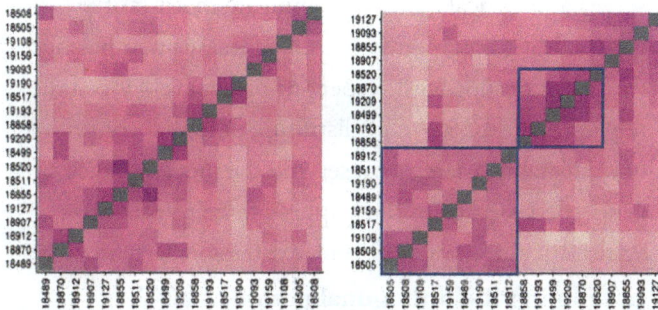

Using the specific data that were analyzed to create this figure – this is a real example - it was later determined that the samples forming the bottom-left cluster in the heatmap had followed a slightly different differentiation pathway toward the terminal cell type compared with the other samples in the study. Complicating matters further, this divergence in the differentiation trajectory occurred only during the middle part of the time course. By the later time points, including the terminal cell type, the differentiation process was consistent across all samples. This variability in timing made it particularly challenging to detect the biological batch effect until the dynamic eQTLs had already been mapped, and this pairwise correlation analysis was performed.

What have we learned from this example of study design?

The examples of standard and dynamic eQTL studies highlight the following key principles:

- **Addressing batch effects through randomization**: In eQTL studies, randomization of samples across batches serves as a critical strategy to mitigate the risk of confounding batch effects. Unlike designs that balance treatment groups with respect to batches, eQTL studies rely on randomization to distribute potential confounders unpredictably, thereby preserving the independence of genotype-based tests. This approach helps minimize systematic differences in gene expression across batches.

- **Leveraging the orthogonal nature of eQTL associations**: The independence of unlinked genotypes across unrelated individuals limits the impact of batch effects in standard eQTL studies. This orthogonality ensures that most proximal genotype-gene expression associations remain unaffected by confounders, even when some batch effects align with specific variation in gene expression levels. The orthogonality of the tests and the resulting robustness are appealing properties of the steady-state regulatory QTL study framework.

- **Emphasizing metadata for reproducibility**: Detailed documentation of metadata, such as collection times, processing steps, and sequencing batch information, is essential. This allows researchers to identify, quantify, and address potential batch effects or other sources of variability during the analysis stage.

- **Recognizing and addressing cryptic biological batch effects in dynamic eQTL studies**: In dynamic eQTL studies, where associations are examined across temporal or spatial perturbations, cryptic biological batch effects can arise due to natural variation in biology. Unlike technical batch effects, these biological confounders cannot always be eliminated through study design. Instead, they require careful *post hoc* examination.

The regulatory QTL study we discussed is an example of a **genetic association study**, a broader type of research that explores how genetic variation is associated with inter-individual differences in phenotypes, including molecular traits, physiological features, and diseases. While we've touched on some important aspects of study design, other factors, like population admixture (briefly mentioned earlier) and environmental exposures, are also important to consider. If you're interested in learning more about these topics, there are many available resources to explore further.

Designing with purpose (V): choices and consequences

The lessons and examples discussed so far were presented in detail to emphasize the complexities of study design and the importance of thoughtful decision-making. With a firm understanding of these challenges and the critical role of study design in ensuring reliable insights, I'd like to end the primer by sharing a few additional examples. These real studies are presented to examine the effects of various decisions made during study design, offering further insight into the significance of factors that might otherwise seem minor. I selected these specific examples because each provides a unique lesson. I will present them briefly, highlighting the key takeaways without revisiting aspects already covered in the primer.

When 'no batches' is still a batch

The first example in this section comes from a study that aimed to understand how transcription factors (TFs) influence gene expression and identify which TF binding events are truly functional, meaning those that lead to measurable changes in gene activity. To achieve this, the researchers performed knockdown experiments, using small RNA molecules to reduce the expression levels of 59 different TFs in cultured human immortalized B cells. If you are unfamiliar with this technique, all you need to know for the purpose of appreciating this example, is that the researchers used a method to decrease, or *knockdown*, the levels of specific TFs in an *in vitro* culture of human B cells.

Each TF was knocked down in a separate sub-culture of the same cell line, meaning the researchers conducted 59 independent experiments, one for each TF. After confirming the knockdowns by measuring both RNA and protein levels, the researchers measured the resulting gene expression levels. Because the knockdown experiment involves certain treatments applied to the cells, it is not appropriate to compare gene expression from the knockdown cells to a completely untreated, or "naïve," cell culture. Instead, the correct comparison is to a control culture that undergoes the same procedures, such as adding RNA molecules and handling the cells, but without actually causing a knockdown of any TF. To achieve this, the researchers introduced into the control cultures random RNA molecules, which do not target or reduce the expression levels of any specific gene.

The techniques used in this study were not novel; however, prior to this work, researchers typically knocked down only a handful of genes in their experiments. This allowed them to culture the controls and all the knockdown cells simultaneously, within a single batch, eliminating the need to account for batch effects. In our example, however, it was not feasible to perform all 59 experiments plus controls in a single batch. To address this, the researchers randomly divided the TF knockdowns into three separate experimental batches and included multiple controls within each batch. Since gene expression comparisons were made between knockdown and control samples, **all comparisons were still performed within a batch**, just as they were performed in earlier experiments with only a handful of gene knockdowns, where everything could be managed in a single batch.

The results, however, were unexpected. Despite randomly dividing the TF knockdowns into batches and ensuring that all gene expression comparisons between knockdowns and controls were performed within the same experimental batch, the researchers detected a clear batch effect. They observed that the gene regulatory responses to individual TF knockdowns were correlated within a batch but uncorrelated across batches. This was surprising because there was no biological similarity among the TFs being studied, and certainly no reason why the gene expression response to the knockdown of randomly chosen TFs within the same batch would be more similar to each other than the regulatory response to TF knockdown across batches. Fortunately, the researchers included a few replicate TF knockdowns across batches. Combined with data from the multiple controls within each batch, which should have been similar across batches as well, they were able to account for the batch effect in their analysis.

The key lesson here is that in case-control setups, even when comparisons are confined to a single batch, unexpected patterns and artificial correlation structures can still arise due to batch effects. This insight underscores the importance of including multiple batches or replicates in these types of experiments. For studies involving only a few TFs and controls performed entirely within a single batch, this raises an important concern: the results of multiple TF knockdowns may have been more correlated than they should have been, due to technical factors. With only a single batch and no additional replicates, it would have been impossible to detect or evaluate such issues.

When unequal sample sizes create the illusion of patterns

The second example comes from a series of studies that explored the evolution of gene expression across species. Understanding how gene regulation changes over evolutionary time is important because differences in gene expression are believed to play a key role in both evolutionary adaptations and disease susceptibility. In the comparative functional genomics studies referenced in this example, researchers measured gene expression levels in different tissue samples from humans and several non-human primates, including chimpanzees, orangutans, and rhesus macaques. Their goal was to determine whether gene expression changes gradually over evolutionary time and whether these changes occur under natural selection or are largely neutral (that is, random and not subject to evolutionary constraints).

To investigate this, the researchers compared variation in gene expression within species to differences between species, using the mean squared difference as a measure of overall divergence in gene expression between samples. This measure is commonly used because it ensures all values are positive and provides an intuitive way to quantify the overall magnitude of differences, in this case, regulatory divergence. For those familiar with this field, I note that we will set aside the discussion of the appropriate null evolutionary model, and the gene expression patterns that might be expected under different evolutionary scenarios, as this is not the focus of the lesson from this example.

The researchers observed an interesting pattern: a linear relationship between the mean squared differences in gene expression levels and the time since species diverged. In other words, the mean squared differences in gene expression levels were smaller for comparisons within humans, a tad larger for comparisons between closely related species, and largest for comparisons between more distantly related species. Based on the observation that regulatory divergence increasing linearly with evolutionary distance, the researchers proposed that most gene expression evolves neutrally, without being subjected to strong evolutionary constraints. The researchers suggested that most gene regulatory differences could be explained by random processes.

An evolutionary biologist may wish to discuss this inference and whether the appropriate model was used. However, we are going to focus on something else, which is that **the observed pattern itself was an artifact of the study design**. The problem

157

was that the sample sizes in these comparative studies varied in a way that correlated with the evolutionary distance of the comparisons that were made. The researchers collected data from six donors that were used for within species comparison of gene expression in humans, three from chimpanzees, and only one each from orangutans and rhesus macaques. The imbalance of sample size in the study design was not random; it led to progressively smaller sample sizes for increasingly distant evolutionary comparisons. This feature of the design introduced a statistical bias.

When comparing any two groups, the squared difference between their sample means becomes less reliable as an estimate of the true population difference when the sample size decreases. Smaller sample sizes lead to greater variability in the estimates, making them less stable. In this study, the comparison among humans had the largest sample size, and the sample size gradually decreased for comparisons involving more distantly related species. Given this design, even if there were no true differences in mean gene expression, either within or between species, the mean squared differences between the samples would still show a positive correlation with the decreasing sample size and, consequently, with evolutionary distance as well.

This kind of data structure, where correlations between different types of samples progressively and consistently decrease with increasing distance, is not unique to comparative genomic studies. Similar patterns appear in many contexts, including population genetic data, experimental evolution studies, and ecological research. In population genomics, there is a long history of misinterpreting analytical patterns that arise from this "isolation by distance" structure, where the observed trend is intrinsic to the data itself. This example highlights how an isolation-by-distance feature of the study design itself, rather than the analysis, can produce intrinsic artifacts that can be misinterpreted as meaningful biological signal.

We often consider sample size primarily in the context of a study's statistical power. However, **the key lesson from this example is that unequal sample sizes can introduce systematic biases that influence results in unexpected ways.** When comparing groups, particularly across categories that differ in their degree of isolation, it is important to balance sample sizes as much as possible to ensure that the conclusions reflect true patterns rather than artifacts of the study design.

Down the wrong path: when intuition leads science astray

The next example comes from studies investigating the genetic basis of obesity. Genome-wide association studies (GWAS) identified genetic variants in a region of the genome near the *FTO* gene that were strongly associated with body weight and obesity in humans. These obesity-associated loci were not located within the *FTO* gene itself but slightly upstream of it. Given their genomic position, it was intuitively assumed that these loci serve as regulatory elements of the *FTO* gene, suggesting a link between its regulation and obesity in humans.

To test this hypothesis, mouse models were used to perturb the regulation of the human *FTO* gene's ortholog and study the resulting effects. When animals with a perturbed *FTO* gene showed differences in weight compared to wild-type mice, this observation appeared to support the model linking *FTO* regulation to obesity. This finding generated considerable excitement in the field, prompting several pharmaceutical companies to focus on the *FTO* gene and related pathways, with the intention of finding new targets for obesity-related therapeutics.

An independent researcher, not involved with the original GWAS or the follow-up mouse experiments, raised some doubts. This researcher noted, based on existing data, that experiments in mice often result in changes in weight. An analysis of published studies and databases revealed that about one-third of genetic or regulatory manipulations in mice lead to weight changes. In many cases, these weight changes were not the main focus of the study and were therefore not highlighted as a primary phenotype, but they were present. This observation cast doubt on the specific connection between *FTO* and obesity, because the weight changes could simply reflect a relatively common outcome of genetic manipulations in mice.

The skeptical researcher decided to conduct targeted experiments using various genomic techniques to identify which gene was actually being regulated by the loci associated with obesity. The results revealed that, although the obesity-associated loci are located upstream and near the *FTO* gene, they play a regulatory role in activating a different gene, *IRX3*, which is also nearby but located farther away. The research team demonstrated that the regulatory interaction between the locus of interest and *IRX3* is conserved across multiple species, not just mice.

The research team conducted additional experiments and provided further evidence for their findings, using multiple experimental approaches, platforms, and functional and phenotypic studies. Their results consistently showed that *IRX3*, not *FTO*, is the gene being regulated by the locus associated with obesity in humans. Their work is a true *tour de force* of careful experimental design and comprehensive functional analysis, and it is well worth reading (the citation is provided at the end of the book).

For our purposes, this example offers several key lessons. *First*, while it is widely understood that correlation does not imply causation, we are still prone to falling into this trap when the inferred causal relationship appears intuitive. This example highlights how problematic and misleading such assumptions can be. *Second*, contextualizing results is essential. To properly evaluate an observed pattern, one must design an effective study that enables an evaluation of how unusual the observation is within a broader context. Without this, it becomes difficult to judge whether the pattern reflects a meaningful discovery or something that could occur by chance. *Third*, although the initial inference that the locus regulated *FTO* was intuitively appealing and accepted with relatively little evidence, the research team that challenged this model had to generate a substantial body of evidence before their alternative explanation was accepted. This reflects a common dynamic in science: it is often far more difficult to challenge and overturn an existing observation than to propose the original one. This dynamic further emphasizes why flawed studies can have such a significant detrimental impact on scientific progress.

Comparing datasets: challenges related to causal inference

The last example does not strictly provide a lesson in study design but rather offers a broader understanding of statistical inference, which is highly relevant to study design. It is not an example of a specific study but instead represents a class of studies.

Functional genomics studies are often designed to infer causal relationships. However, as we have discussed multiple times, these studies can only generate causal hypotheses, which must be tested in follow-up experiments (the *FTO* / *IRX3* example we just discussed clearly illustrated this point!). I like to describe the goal of genomic studies aimed at causal inference as building a *circumstantial case* for a causal relationship. The inferred causality may be consistent with all observations, but it remains putative until direct evidence is obtained. To build a circumstantial case for causality using functional genomics data, we often need multiple lines of evidence. For example, if we wish to infer a causal relationship between transcription factor binding and target gene activity, we might combine results from an assay that provides us with information on where the transcription factor binds, with results from an assay that measures changes in gene expression following the knockdown of the same transcription factor.

On their own, these data provide limited insight into causality. Transcription factors can bind many genomic regions without influencing gene expression, and knocking down a transcription factor can lead to changes in the expression of many genes that are not directly regulated by it. However, when combined, these data can strengthen causal inference. If we observe that a transcription factor binds near regulatory elements associated with a gene, and knocking down the transcription factor results in a change to the gene's expression, the combination of these observations strengthens the case for a causal relationship. While still circumstantial, this overlap suggests a link between transcription factor binding and gene regulation.

To illustrate another approach, we might want to identify distant regulatory elements and their target genes. We can use *Hi-C*, which is a technique that enables mapping of physical interactions between different parts of the genome, to identify distant regulatory regions (like enhancers) that physically interact with local promoters. The *Hi-C* data alone suggests that these genomic regions might be physically connected, but it does not establish regulatory activity. To test whether this physical interaction is linked to gene regulation, we can use eQTL mapping, for example, to identify genetic variants in distant enhancers that are associated with variation in gene expression levels. If the combined data reveal physical interactions between distant enhancers and the promoters of genes whose expression differs depending on the genotype at the enhancer (an eQTL), this strengthens the circumstantial case for a causal relationship.

The common theme in these two examples is that drawing putative causal inferences from functional genomics data often relies on comparing and identifying overlaps across different datasets. The challenge in doing so lies in our limited power to detect patterns of interest, such as transcription factor binding, changes in gene expression, or differences in chromatin accessibility. This limitation arises because no experiment or analysis can perfectly capture all true signals in the data. As a result, comparisons of "significant observations" between data sets are susceptible to false negatives. In this context, a false negative occurs when a true signal is detected in one data set but missed in another. This can create the appearance of a difference between the two data sets, when in reality, the signal was present in both yet was undetected in only one data set due to limited power.

To understand this intuitively, consider the issue of interpreting statistical significance using an arbitrary cutoff. For instance, an observation with a p-value of 0.049 is often considered significant, while one with a p-value of 0.051 is not. However, in practical terms, the difference between these two results is negligible. A p-value of 0.049 means the observed result would occur by chance approximately 1 in 20.4 times, while a p-value of 0.051 corresponds to a chance occurrence of about 1 in 19.6 times. Despite this minimal difference, one observation meets the arbitrary common threshold for significance, while the other does not, highlighting the limitations of rigid cutoffs.

When analyzing a single data set, it is common to rely on a single arbitrary threshold, such as a *p-value* cutoff, to control for false positives. This approach reflects a preference for avoiding incorrect claims about unusual findings (false positives), over the risk of missing such findings by mistakenly classifying them as typical (false negatives). However, when comparing results across multiple data sets, this reliance on arbitrary thresholds can introduce a different problem: false negatives within each data set can create the illusion of differences between them. In reality, the signal may exist in both data sets but fail to meet the threshold in one due to limited power, leading to false conclusions – false positives - regarding discrepancies.

To address this, we need to use statistical approaches that account for incomplete power. These approaches recognize that the data we observe are noisy, power is incomplete, and some true signals may fall just below the arbitrary threshold of detection. Instead of treating each data set independently, we can combine evidence across multiple sources, increasing our ability to detect patterns and reducing the chance of drawing incorrect conclusions from incomplete observations.

Appreciating this issue is also important for study design because it helps clarify and prioritize our goals, particularly in determining which types of inferences are most important. Designing an experiment to guard against false positives requires a different approach than one focused on minimizing false negatives. While we always aim to optimize both, the balance between these tradeoffs can and should vary depending on the specific context and objectives of the study.

Concluding thoughts

This primer provides a foundational introduction to the principles of effective study design for functional genomics, offering guidance on navigating the complexities of experimental planning in a field where nearly any deviation from the expected distribution can be considered an interesting result.

While the primer provides a structured framework and illustrative examples, it is important to acknowledge that the scope of study design extends far beyond what is covered here. The field includes nuanced statistical methodologies, ethical dimensions, and innovative approaches across various types of data. For those interested in a deeper understanding, the resources listed at the end of this text provide initial avenues for further exploration.

If there is one key takeaway from this primer, it is the necessity of being deliberate and thoughtful at every step of the study design process. Study design is not just about generating data - it is about generating data that are meaningful, interpretable, and capable of addressing the intended research question. Mistakes at this stage can have far-reaching consequences, often rendering datasets incomplete, biased, or even unusable. While it is true that not all questions in study design will have a single correct answer, it is equally true that some answers are clearly wrong. Decisions that introduce systematic confounders, misalign study goals with technology, or fail to anticipate potential biases can lead to data that misrepresent reality or fail to answer the research question altogether. Unfortunately, there are many published studies whose conclusions are fundamentally undermined by these types of errors.

Designing an effective study is not about finding the perfect approach; it is about choosing a path that minimizes weaknesses and maximizes reliability and reproducibility. It requires clarity of purpose: What is the specific question or hypothesis? What are the potential sources of bias or variability? And how will the study account for them? Clarity also involves acknowledging and managing trade-offs. As we discussed, there is often a balance between precision and generalizability. Making these decisions requires not only technical knowledge but also a commitment to aligning every aspect of the design with the goals of the study.

In addition to deliberate planning, effective study design also involves heuristic and global thinking. Researchers must think beyond the immediate scope of their study, considering the broader context of their work and anticipating how decisions at the design stage will influence downstream analyses and interpretations. This requires looking for gaps, questioning assumptions, and exploring alternative approaches. While some variability or uncertainty is inevitable in any experiment, robust design minimizes these challenges and ensures that the data collected are capable of yielding meaningful conclusions.

Feedback and collaboration are invaluable, bringing diverse perspectives that can uncover blind spots and strengthen study design. Constructive critique and learning from past mistakes, both personal and within the scientific community, are essential for improvement. Poorly designed studies are not just missed opportunities; they can propagate misinformation and mislead future research efforts. They are also leading to a waste of time and resources. By committing to thoughtful design, researchers contribute not only to their own projects but also to the integrity of the field at large.

Finally, science is iterative, and study design is no exception. Pilot studies are one example of how iteration can refine experimental approaches, providing critical insights before large-scale efforts are undertaken. Adaptability is essential, as even the best-planned studies can encounter unforeseen challenges. Remaining flexible and open to adjusting the design in response to

166

preliminary findings or new constraints ensures that the research stays on course without compromising its goals.

In closing, this primer provides an intuitive framework for approaching study design, emphasizing the importance of clarity, rigor, and deliberate planning. It is only a starting point. Mastery of study design is an ongoing process, requiring continuous learning, practice, and collaboration. By adopting a thoughtful approach to experimental planning, researchers can ensure that their work not only addresses immediate questions but also contributes meaningfully to the collective knowledge of the field.

Appendix A: suggested design for time course studies

When conducting studies with multiple classes of samples, such as different treatments or time points, the goal is often to compare data directly across these classes. For example, a study might aim to compare treatment effects to one another rather than simply comparing each treatment to a control. In such studies, it is important to employ a design that prevents confounding sample processing batches with the different classes of samples.

The concepts we covered in this primer should provide you with the foundation needed to design such an experiment. If you have not yet read other texts on study design and do not have formal education in statistics, you might be unfamiliar with the standard terminology for the design feature you are likely to implement, which is called *blocking*. However, even without knowing the specific terminology, the principles we have covered and discussed will guide you in achieving a blocking study design.

For example, in a study aimed at comparing the regulatory response across multiple treatments, you now understand the importance of separating the sample processing batch from the treatment assignment. In this context, the processing batch serves as the block, with the different treatments either balanced or randomized within each block. The purpose of this *appendix* is to outline the corresponding blocking design for time course studies. In principle, the design should follow a similar approach to the one we employ for different treatments, incorporating *blocking* of the sample processing batch with respect to the sampling times during the time course. However, implementing the blocking design in the context of time course studies requires a somewhat unconventional experimental approach.

The challenge with time course studies is that blocking the sample processing batch becomes less straightforward because the variable class is time. In most time course experiments, samples within a batch are typically processed on the same schedule, resulting in synchronized sampling time points. For instance, in a time course study involving 10 cell cultures that are all initiated simultaneously, temporal sampling from all cultures will occur at the same time for each time point. This creates a confounding effect between the time points and the sample processing batches.

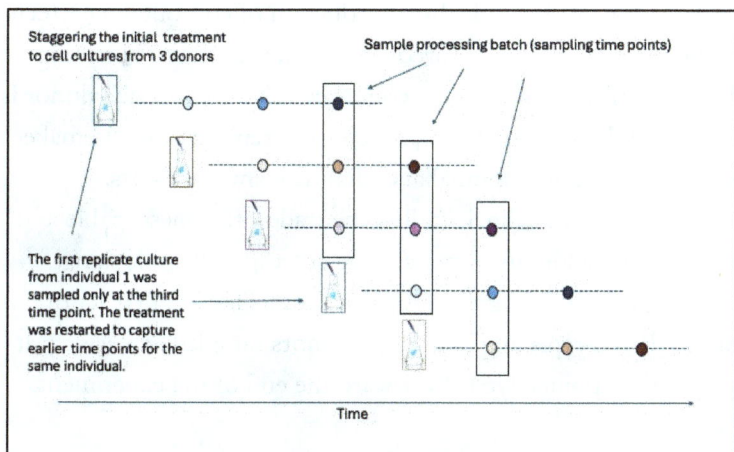

The figure on this page illustrates how this confounding effect can be resolved through an effective blocking design. It demonstrates a time course study design using cell cultures from three donors, each sampled at three different time points. The sampling time points are represented by circles, with shades of color transitioning from light to dark to indicate earlier to later time points, respectively. In practice, each culture in this figure can represent a larger number of donors or samples, depending on the logistical feasibility of conducting multiple experiments simultaneously. By staggering samples along the time course experiment, we effectively separate sampling times from sample processing batch. A distinctive aspect of this design is the need to repeat the experiment for certain individuals to capture their earlier time points within a consistent sample processing batch.

This design, as shown, presents a trade-off as it effectively eliminates the association between processing batches and sampling times but introduces a non-overlapping technical time course replicate for a subset of the samples (for the first and second individuals in this example). Although this approach may increase variance, it avoids the introduction of confounding effects, and therefore preferable. Moreover, in many contexts, obtaining technical replicates of the time course experiment for each donor is desirable, and applying this approach to all replicates would make the staggering design optimal and eliminate any trade-offs.

In *in vivo* longitudinal studies, the trade-off is inescapable and involves using additional donors. The corresponding blocking design requires skipping the early time points for the first set of individuals and replacing them with donors sampled exclusively at the early time points, typically toward the end of the experiment.

Appendix B: from correlations to specific hypotheses

As we mentioned at the beginning of the primer, functional genomics studies offer unparalleled freedom to explore, providing an exceptional opportunity to uncover patterns and relationships across diverse datasets. This freedom is not constrained by predefined hypotheses, allowing researchers to investigate any correlation that emerges. The process of exploration and generating hypotheses is not part of study design, but it is worth discussing briefly here, as it helps clarify a common goal of functional genomics studies, namely, identifying interesting patterns that guide further research.

The exploratory phase in functional genomics is different from data analysis in hypothesis-driven research. Exploration can be characterized by an unrestricted examination of correlations, where the objective is not to confirm specific ideas but to discover intriguing patterns that could inspire new hypotheses. To achieve this, researchers often create comprehensive integrated datasets that combine information from the study itself with external sources, including functional genomics annotations, molecular interaction networks, and evolutionary conservation data. For example, integrating gene expression profiles, chromatin accessibility data, and metabolomic data might reveal a correlation pattern whereby genes within specific metabolic pathways exhibit coordinated regulatory changes under different conditions, potentially driven by changes in binding of specific transcription factors. This discovery could point to a broader regulatory mechanism influencing these pathways, which could then be further investigated.

The lack of constraints during the exploration phase is liberating. There is no need to adjust for multiple testing or worry about false positives, as the goal is not validation but inspiration. This freedom allows researchers to cast an exceptionally wide net, identifying correlations that might otherwise go unnoticed. It is important to clarify, however, that this approach is not a substitute for rigorous statistical analysis in contexts where it is appropriate. For instance, when analyzing data to identify differentially expressed genes, it is essential to apply standard statistical methods, including corrections for multiple testing, to ensure the results are both robust and reproducible. Instead, what I am emphasizing is that beyond these essential, structured analyses, functional genomics datasets offer a unique and vast landscape for discovery.

Exploration is not about guaranteeing that every pattern uncovered will lead to a breakthrough; many leads will prove to be false. This is not a limitation but an essential aspect of the process, which must be acknowledged explicitly. The exploratory phase lays the groundwork for discovery by enabling the formulation of hypotheses that can then be tested in follow-up experiments.

Not every correlation observed during the exploratory phase leads directly to the formulation of a hypothesis or the design of a dedicated new experiment. Instead, researchers refine potential hypotheses by gradually building circumstantial cases. This process involves leveraging known causal relationships to strengthen confidence in the significance of observed patterns and seeking repeated evidence across related explorations. For instance, correlations that align with established biological mechanisms or consistently appear in similar contexts can provide compelling support for their relevance. Through iterative examination and integration of findings from multiple studies, researchers identify patterns that stand out as particularly intriguing, ultimately justifying the formulation of a new hypotheses and the design of targeted hypothesis-driven experiments for validation.

By embracing the exploratory phase as a vital component of the scientific process, researchers can venture into uncharted territories, uncovering new directions and generating insights that drive further inquiry. This unrestricted exploration is the foundation of discovery, enabling us to ask questions we might not have thought to pose and laying the groundwork for meaningful scientific advances.

Additional reading

This is not an exhaustive list but a curated selection of references to complement the material in this primer. Aligned with the topics discussed, these references primarily focus on aspects of study design rather than data analysis

Study design in general

Ruxton GD, Colegrave N. **Experimental Design for the Life Sciences**. 4th ed. Oxford University Press; 2016. *This book provides a comprehensive introduction to experimental design tailored for life science researchers. It covers essential principles and practical strategies to construct effective and valid experiments. Accessible to advanced undergraduates, graduate students, and early-career researchers with a foundational understanding of biological sciences.*

Marder MP. **Research Methods for Science**. Cambridge University Press; 2011. *This text explores a broad spectrum of research methodologies across the sciences, emphasizing critical thinking and robust study design. It includes clear examples and explanations, making it suitable for advanced undergraduates, graduate students, and interdisciplinary researchers new to study design.*

Luo J, Wu X, Cheng Y, Chen G, Wang J, Song X. **Expression quantitative trait locus studies in the era of single-cell omics**. Front Genet. 2023;14:1182579. This review discusses how expression quantitative trait locus (eQTL) studies are adapting to the advent of single-cell technologies. It provides an overview of study design challenges and opportunities in this rapidly evolving field, suitable for graduate students, bioinformaticians, and genomics researchers.

Kelley JL, Gilad Y. **Effective study design for comparative functional genomics**. Nat Rev Genet. 2020 Jul;21(7):385-386. *This article outlines best practices for designing studies in comparative functional genomics, emphasizing reproducibility and robustness. Suitable for researchers with a working knowledge of genomics.*

Srinagesh K. **The Principles of Experimental Research**. Butterworth-Heinemann; 2006. *This book delves into the theoretical underpinnings of experimental research, covering a wide array of topics from basic principles to complex designs. It's ideal for graduate students and professionals in science and engineering seeking a deeper theoretical understanding of experimental approaches.*

Batch effects in genomics studies

Yu Y, Mai Y, Zheng Y, Shi L. **Assessing and mitigating batch effects in large-scale omics studies**. Genome Biol. 2024;25:254. *This article provides a detailed discussion of batch effects, their impact on large-scale omics studies, and methods for their mitigation. Suitable for genomics researchers and bioinformaticians with intermediate knowledge of data processing and statistical modeling.*

Price EM, Robinson WP. **Adjusting for batch effects in DNA methylation microarray data, a lesson learned.** Front Genet. 2018 Mar 23;9:83. *This study highlights the impact of batch effects in DNA methylation microarray experiments and emphasizes the importance of proper sample randomization to mitigate these effects.*

Goh WWB, Wang W, Wong L. **Why batch effects matter in omics data, and how to avoid them.** Trends Biotechnol. 2017 May;35(5):498-507. *This review highlights the significance of batch effects in omics research, explaining their origins and solutions. It's written in an accessible manner, suitable for graduate students, early-career researchers, and those new to omics studies.*

Normalization

Qin S, Kim J, Arafat D, Gibson G. **Effect of normalization on statistical and biological interpretation of gene expression profiles.** Front Genet. 2013 May 31;3:160. *This study evaluates the impact of nine different normalization strategies on gene expression datasets, focusing on their effects on statistical and biological inference. It highlights how methods like mean centering and quantile normalization can influence the detection of biological covariates.*

Johnson KA, Krishnan A. **Robust normalization and transformation techniques for constructing gene coexpression networks from RNA-seq data.** Genome Biol. 2022;23:1. *This study provides a comprehensive benchmarking of 36 workflows for constructing gene co-expression networks from RNA-seq data, focusing on normalization and network transformation methods. It is suitable for researchers in computational biology, bioinformatics, and genomics with experience in RNA-seq data processing.*

Lun ATL, Bach K, Marioni JC. **Pooling across cells to normalize single-cell RNA sequencing data with many zero counts.** Genome Biol. 2016 Apr 27;17:75. *This paper introduces a method for normalizing single-cell RNA-seq data by pooling expression values across cells to address high dropout rates and technical noise. The authors show that their* *deconvolution-based approach improves accuracy compared to other methods. This reference is suitable for researchers with background in computational biology.*

Studies we mentioned as examples in this primer

Martin-Magniette ML, Aubert J, Cabannes E, Daudin JJ. **Evaluation of the gene-specific dye bias in cDNA microarray experiments.** Bioinformatics. 2005 May 1;21(9):1995-2000. *This study investigates dye bias in cDNA microarray experiments, proposing evaluation methods to correct for these biases. It is suitable for researchers familiar with microarray technologies and statistical bias correction, who are interested in some of the history of the field...*

Cusanovich DA, Pavlovic B, Pritchard JK, Gilad Y. **The functional consequences of variation in transcription factor binding.** PLoS Genet. 2014 Mar 6;10(3):e1004226. *This study examines the functional impact of transcription factor (TF) binding by combining knockdown experiments for 59 TFs with ChIP-seq and DNase-seq data. It reveals that only a subset of TF binding events directly influence gene expression, with functional binding enriched at enhancers and binding sites that are inferred to have stronger affinity.*

Lin S, Lin Y, Nery JR, Urich MA, Breschi A, Davis CA, Dobin A, Zaleski C, Beer MA, Chapman WC, Gingeras TR, Ecker JR, Snyder MP. **Comparison of the transcriptional landscapes between human and mouse tissues.** Proc Natl Acad Sci U S A. 2014 Dec 2;111(48):17224-17229. *This is one of the mouse-ENCODE studies we provided as an example.*

Gilad Y, Mizrahi-Man O. **A reanalysis of mouse ENCODE comparative gene expression data.** F1000Res. 2015;4:121. *This is the reanalysis of the mouse-ENCODE data, which identified the batch effect related to sequencing lane assignment.*

Pozhitkov AE, Neme R, Domazet-Lošo T, Leroux BG, Soni S, Tautz D, Noble PA. **Tracing the dynamics of gene transcripts after organismal death.** Open Biol. 2017 Jan;7(1):160267. doi:10.1098/rsob.160267. *This study investigates the post-mortem dynamics of gene expression, proposing that transcripts remain active and exhibit specific patterns even after organismal death. We discussed this study in the context of empirical normalization.*

Gilad Y, Rifkin SA, Bertone P, Gerstein M, White KP. **Multi-species microarrays reveal the effect of sequence divergence on gene expression profiles.** Genome Res. 2005 May;15(5):674-680. *This study used a microarray designed with probes that match different primate species to characterize the impact of sequence mismatches on hybridization intensity and demonstrate the cryptic effects of normalization on the reported results of comparative gene expressions studies using single-species arrays.*

Harper KN, Peters BA, Gamble MV. **Batch effects and pathway analysis: two potential perils in cancer studies involving DNA methylation array analysis.** Cancer Epidemiol Biomarkers Prev. 2013 Jul;22(7):1052-1060. *This article examines the influence of batch effects on DNA methylation studies and their potential to skew pathway analysis in cancer research. It emphasizes the need for careful experimental design and statistical controls to ensure reliable conclusions.*

Smemo S, Tena JJ, Kim KH, Gamazon ER, Sakabe NJ, Gómez-Marín C, Aneas I, Credidio FL, Sobreira DR, Wasserman NF, Lee JH, Puviindran V, Tam D, Shen M, Son JE, Vakili NA, Sung HK, Naranjo S, Acemel RD, Manzanares M, Nagy A, Cox NJ, Hui CC, Gomez-Skarmeta JL, Nóbrega MA. **Obesity-associated variants within FTO form long-range functional connections with IRX3**. Nature. 2014 Mar 20;507(7492):371-375. *This study found that obesity associated loci near the FTO gene are actually regulating a different, more distant gene, IRX3. It highlights the importance of conducting careful and specific experiments to follow up and test causal inference based on genomic data.*

Gilad Y, Oshlack A, Rifkin SA. **Natural selection on gene expression.** Trends Genet. 2006 Aug;22(8):456-461. *This older review article discusses approaches for studying natural selection on gene regulation. It also highlights the artifact arising from progressively smaller sample sizes in distantly related evolutionary comparisons.*

Maceda I, Lao O. **Analysis of the batch effect due to sequencing center in population statistics quantifying rare events in the 1000 Genomes Project.** Genes (Basel). 2022 Jan 5;13(1):44. *This study investigates how sequencing center-specific batch effects influence the quantification of rare genetic events in population-scale datasets, using the 1000 Genomes Project as a case study. It highlights the importance of addressing batch effects in large-scale sequencing projects to ensure accurate population genetic inferences.*

Pai AA, Cain CE, Mizrahi-Man O, De Leon S, Lewellen N, Veyrieras JB, Degner JF, Gaffney DJ, Pickrell JK, Stephens M, Pritchard JK, Gilad Y. **The contribution of RNA decay quantitative trait loci to inter-individual variation in steady-state gene expression levels.** PLoS Genet. 2012;8(10):e1003000. *This paper explores the role of RNA decay in inter-individual gene expression variation. We discussed this study in the context of empirical normalization.*

Power analysis

Although we did not cover power analysis in this primer, we have included a few references to help you begin exploring this important topic.

Jeon H, Xie J, Jeon Y, Jung KJ, Gupta A, Chang W, Chung D. **Statistical power analysis for designing bulk, single-cell, and spatial transcriptomics experiments: review, tutorial, and perspectives.** Biomolecules. 2023;13(2):221. *This review and tutorial focus on statistical power analysis in transcriptomics, offering guidelines for bulk, single-cell, and spatial experiments. Suitable for those with background in bioinformatics and/or biostatistics*

Cao C, Ding B, Li Q, Kwok D, Wu J, Long Q. **Power analysis of transcriptome-wide association study: Implications for practical protocol choice.** PLoS Genet. 2021 Feb 26;17(2):e1009405. *This article addresses power considerations in transcriptome-wide association studies, offering practical insights for researchers designing such studies. Suitable for geneticists and computational biologists familiar with association study methodologies.*

Ledolter J, Kardon RH. **Focus on data: statistical design of experiments and sample size selection using power analysis.** Invest Ophthalmol Vis Sci. 2020 Jul 1;61(8):11. *This paper presents a case study on statistical design and sample size determination using power analysis, with a focus on ophthalmology research. Accessible to researchers and students with foundational knowledge in statistics and experimental design.*

www.ingramcontent.com/pod-product-compliance
Lightning Source LLC
Chambersburg PA
CBHW071551200326
41519CB00021BB/6699